AGR 7519

THE DYNAMICS OF

DIGITAL EXCITATION

THE DYNAMICS OF

DIGITAL EXCITATION

by

Masakazu Shoji
Computing Science Research Center
Lucent Bell Laboratories

KLUWER ACADEMIC PUBLISHERS
Boston / Dordrecht / London

Distributors for North America:
Kluwer Academic Publishers
101 Philip Drive
Assinippi Park
Norwell, Massachusetts 02061 USA

Distributors for all other countries:
Kluwer Academic Publishers Group
Distribution Centre
Post Office Box 322
3300 AH Dordrecht, THE NETHERLANDS

Library of Congress Cataloging-in-Publication Data

A C.I.P. Catalogue record for this book is available
from the Library of Congress.

Copyright © 1998 by Kluwer Academic Publishers

All rights reserved. No part of this publication may be reproduced, stored in a retrieval system or transmitted in any form or by any means, mechanical, photocopying, recording, or otherwise, without the prior written permission of the publisher, Kluwer Academic Publishers, 101 Philip Drive, Assinippi Park, Norwell, Massachusetts 02061

Printed on acid-free paper.

Printed in the United States of America

THE DYNAMICS OF

DIGITAL EXCITATION

This Book is Dedicated to

Contents

1 Propagation of Digital Excitation in the Gate Field

1.01	Introduction	1
1.02	Examples of the Gate Field	3
1.03	The Vector Gate Field	8
1.04	Energy Transfer in Gate Field	10
1.05	CMOS Inverter Switching Process	20
1.06	The Velocity of the Propagation of Excitation	39
1.07	An Equation of the Motion of Excitation	45
1.08	Node Waveform of Logic Circuits	49
1.09	Logic Threshold Voltage and Gate Delay Time	52
1.10	Nonmonotonous Node-Switching Voltage Waveforms	56
1.11	The Strange Consequences of the Classical Delay-Time Definition	59
1.12	The Phase Transition of the Gate Field	64
1.13	The Miller Effect in the Gate Field	67
1.14	Feedforward Excitation Transmission	73
1.15	The Gate Field of a Negative-Resistance Diode	79

2 Quantum Mechanics of Digital Excitation

2.01	Introduction	83
2.02	Elementary and Composite Excitation	84
2.03	Finite and Infinite Energy Associated with Excitation	86
2.04	An Eigenvalue Problem in the Gate Field	89
2.05	The Eigensolution of a Gate-Field Waveform	94
2.06	Gate-Field Variable Measurements	98
2.07	Latch Circuit for Boolean-Level Determination	101
2.08	The Decision Threshold	105

2.09	The Probabilistic Interpretation of Boolean Level	109
2.10	Metastability in Observation	117
2.11	Propagation of Excitation through a Nonuniform Field	121
2.12	The Tunnel Effect of Digital Excitation	124
2.13	Ambiguity in the Cause and Effect Relationship	130
2.14	Valid Delay-Time Measurement of the Digital Circuit	136
2.15	The Quantum-Mechanical Delay Definition	141
2.16	Design Guidelines for Ultrafast Circuits	147
2.17	Natural Decay of Composite Excitation	159
2.18	A Theory of the Decay of Isolated Pulses	170
2.19	Mass of Digital Excitation	178
2.20	The Dynamics of Digital Excitation in Closed Path	184

3 The Macrodynamics of Digital Excitation

3.01	Introduction	189
3.02	Quantum States	191
3.03	Bohr's Correspondence Principle	196
3.04	States of Nodes and Circuits	201
3.05	The Capability of a Circuit to Store Information	205
3.06	Information Stored in a Ring	210
3.07	Extraction of the Features of Data Pattern	218
3.08	Digital Excitation in a Closed Path	220
3.09	Multiple Ringoscillators	223
3.10	A General Observation of Ringoscillator Dynamics	235
3.11	Modes of Oscillation	242
3.12	A State-Space Representation of Ringoscillator	248
3.13	The Practical Significance of Ringoscillator Logic	256
3.14	An Asynchronous Multiloop Ringoscillator	260
3.15	The Precision of an FET Model and Simulator	265
3.16	Conclusion	270
3.17	The Future Direction of Digital-Circuit Research	271

Preface

The electronic circuit is a proud child of twentieth century natural science. In a hundred short years it has developed to the point that it now enhances nearly every aspect of human life. Yet our basic understanding of electronic-circuit operation, *electronic – circuittheory*, has not made significant progress during the semiconductor industry's explosive growth from 1950s to the present. This is because the electronic circuit has never been considered to be a challenging research subject by physicists. Linear passive circuit theory was established by the late 1940s. After the advent of the semiconductor electron devices, the interest of the technical community shifted away from circuit theory. Twenty years later, when integrated circuit technology began an explosive growth, circuit theory was again left behind in the shadow of rapidly progressing computer-aided design (CAD) technology.

The present majority view is that electronic-circuit theory stands in a subordinate position to CAD and to device-processing technology. In 1950s and 1960s, several new semiconductor devices were invented every year, and each new device seemed to have some interesting fundamental physical mechanisms that appeared worth investigating. Compared to attractive device physics, the problems of the semiconductor device circuit appeared less sophisticated and less attractive. Bright minds of the time drifted away from circuit theory to electron-device physics. After thirty years only one type of semiconductor device, the electron triode with several variations survived, whereas hundreds of them went into oblivion. When CAD technology began to expand in the 1970s, its new features dazzled the technical community. Again, circuit theory appeared obsolete, unattractive, and not promising for younger physicists. Today many CAD programs are written using quite shaky physical models of circuits on silicon.

These are the historical reasons why integrated-circuit (IC) technology has never been supported by adequate circuit theory. With this history in mind, I have to ask a question: Is a circuit-theoretical problem really less sophisticated and less attractive than any other problem of physics? When I first asked myself this question, I was an aspiring semiconductor device physicist in about 1965 the answer seemed to be yes. After thirty years as a practicing integrated-circuit designer, however, my answer is no. Historical context is important. I have published three books over the last ten years to promote my revised point of view. I tried to persuade the readers of my books, but the continued emergence of new and interesting problems has given me the confidence to make an even more ambitious claim, that electronic circuit must exist in nature as a fundamental object. Based on this conviction, I wish to explore another well-defined problem in this book. The logical relationship between digital-circuit theory and mechanics, including quantum mechanics. A digital circuit is a model of a quantum-mechanical world, based on Einstein, Podorsky, and Rosen's hidden variable interpretation. This new viewpoint has many practical applications to integrated-circuit engineering. I have found many applications to my work as an IC designer, and I wish to share them with my fellow IC designers.

There are several philosophical reasons for making this inquiry. The physical basis of computation has not yet been well understood. For instance, the fundamental question about how much energy is required to execute a logic operation has not been answered adequately. Fundamental issues of this sort cannot be resolved by considering practical design problems alone. We need to consider what a digital signal is. A *digital signal* is an excitation launched to a saturable circuit chain that works as a *field* in the physicist's terms. Since the excitation is an identifiable object, why cannot it be described by an equation of motion of a particle? An excitation has many ordinary characteristics, but it has too many peculiarities to be considered to be an object of the classical mechanics domain. To understand this object, quantum mechanics must be used.

Digital excitation is created and destroyed in the field. Elementary quantum mechanics is not able to deal with such a phenomenon, quantum field theory is required. Some of these theoretical subjects may not interest readers, who want to focus solely on practical issues. Readers who are interested only in IC engineering might want to skip the middle of Section 1.04 and Sections 2.19, 2.20, and 3.03.

In recent years, a barrier that limited the territory of natural science, an inquiry into the human mind, has been slowly but steadily lifting. Several influential authors relate its operation to quantum mechanics. Yet a neuronic logic gate in the brain is definitely a macroscopic object. If a certain quantum-mechanical process in the brain creates the human mind, quantum mechanics must be operative not at microscopic levels but in macroscopic hardware that behaves like microscopic objects. If the views of influential authors are rephrased in this way, the problem becomes suddenly quite attractive to very large scale integrated circuit (VLSI) designers, who constantly struggle to create more and more intelligent ICs. From this viewpoint, the apparent similarity between digital circuits and quantum-mechanical objects requires an in-depth investigation. From these reasons, in this book I study the properties of a digital excitation as an object of classical and quantum mechanics.

This new approach provides a new viewpoint for understanding something more immediately important than the human mind. The idea of analyzing a digital circuit as a whole, and not from the parts, will provide a new algorithm for CAD. The theory clarifies the digital-circuit delay times that have always puzzled circuit designers. Digital delay time is imprecisely defined, and such a fundamental issue never fails to have a large impact on the immediate circuit-design problem. The delay-time definition problem has never been recognized as a fundamental issue, but in high-speed ICs, the problem exists everywhere. I am using the theory of this book in ultrahigh-speed complementary metal-oxide-semiconductor (CMOS) circuit design, and I have a much better understanding and a much better delay-time estimate for design verification than the conventional method would have given me. Furthermore as the digital-circuit delay time is defined better, better guidelines for designing a high speed digital circuit emerge, and that is closely connected to the quantum-mechanical tunnel effect. I am certain that in-depth study of digital circuits always leads us to something new and useful.

Acknowledgments

This book was written with the encouragement and valuable comments of Professor E. Friedman of the University of Rochester, Rochester, New York; Professor S. M. Kang of the University of Illinois, Urbana, Illinois; Professor J. Allen of the Massachusetts Institute of Technology and Dr. Manjunath Shamanna of the Somerset Design Center, Motorola, Inc., Austin, Texas. Mr. C. Harris of Kluwer Academic Publishers accepted to publish this highly specialized book. I am grateful for their encouragement and support of my work.

My management and coworkers in the Allentown-Cedarcrest Integrated Circuit Design Center, Lucent Microelectronics, helped me in many ways to conceive and to complete this book. Above all, I am grateful to Mr. G. L. Mowery, and Mr. W. W. Troutman who gave me an opportunity to work there as a consultant, to Messrs M-S Tsay, R. Wozniak, John Kriz and Jim Chlipala, with whom I shared excitement of high performance circuit design, and to Ms. Jyoti Sabnis, with whom I worked on various integrated circuit problems over past 15 years. Mr. T. Thomasco, the center director, and Mr. C. Crawford, Lucent Microelectronics Presedent, warmly supported my technical activities in the Cedarcrest Design Center. I am grateful for all their support of my work.

During the period from 1985 to 1995, the Computing Science Research Center of the AT & T Bell Laboratories consistently supported my activities in the theoretical research of integrated circuits. At the end of December 1995, the scientific organization was dissolved, as a part of the AT & T breakup announced in September 1995. I wish to take this opportunity to thank the generations of managers who supported my activities. The most important was Mr. T. G. Szymanski, who was my superior and who provided constant support, encouragement, and technical suggestions. Messers. B. W. Kernighan, J. H. Condon, A. Aho, R. Sethi, W. Coughran and A. G. Fraser supported me in various ways and gave me encouragements. This book was typeset by the help of Mr. Kernighan. Mr. Fraser hired me to join the respected scientific organization. Professor Adam Heller of the University of Texas at Austin was a research department head of the Bell Laboratories in 1984, and he gave me strong encouragement and a recommendation to move from development to research. In the twilight of my nearly forty years of association with semiconductor technology, I will miss the eleven years of continuous intellectual superheat, and I am grateful for those people who helped me during that time. The main body of this book took its shape within a two-month period immediately following the announcement of Bell Laboratories' breakup in September 1995, as an elegy to the historically great scientific institution. I needed spiritual energy to discover my direction in the confusion of the breakup. During that traumatic winter, I wandered the Maya region of Central America, desperately seeking for inspiration from the spirits of the people who once built a great civilization. The message that reached me was that there exists something much greater than any human institution to which I must dedicate myself. By that message I found my way and a new meaning for human life. That is why this book is dedicated to the two great monarchs and cultural heroes of the classical Maya, King Pacal the Great of Palenque and King Eighteen-Rabbit the Great of Copan. Last, but not the least, I thank my wife, Marika, who joined my pilgrimage to Central America. An accomplished psychoanalyst, she supplied me with much inspiration as I wrote this work, which lies on the remote boundary region of physics and psychology, with understanding of the passion of her husband, who, like the classical Mayas, struggles to build a huge spire in the unexplored and unhospitable jungle region of natural science.

Chapter 1

Propagation of Digital Excitation in the Gate Field

1.01 Introduction

Many compelling similarities among mechanical phenomena, including quantum-mechanical phenomena and digital-circuit transients provide motivation to initiate a research project to examine whether or not there is any deep-level connection among them. The theory of relativity showed that mass is a form of energy. Why, then, can a digital excitation that carries energy not be considered to be a mass point? Yet such a study would only serve to satisfy pure academic curiosity, if that were its only rationale. There are other incentives to go forward with such a study. Since its introduction, the quantum world has remained a fantasy world, accessible only to those armed with logical reasoning and advanced mathematics. The connection between the digital circuit and quantum mechanics provides a new viewpoint. Since the advent of the microprocessor, digital circuits are everywhere. As a model of the quantum world, they can help make the mystery world more accessible. At this point, a basic question must be asked: is the quantum world really unusual or exotic, or it is just a part of nature that happened to escape our attention until recent years due to a lack of familiar examples? I am inclined to believe that the quantum world is really a part of tangible nature and that it can have many varieties. The function of the human mind has been attributed by several influential authors to quantum-mechanical phenomena (Stapp, 1993; Penrose, 1994). Yet the neuronic systems are more concretely described as a special digital circuit than as a quantum-mechanical object. What then are the functional connections between the two?

A familiar question why is the human mind so intelligent? may find an unexpected answer in digital-circuit theory, through the quantum-mechanical viewpoint. The fastest possible digital circuits have a purely quantum-mechanical model, as I show in Sections 2.14 through 2.16. This conclusion does have a practical impact on future high-speed digital circuit design. Thus, the study has both academic and practical significance.

Nature displays physical phenomena in three-dimensional space and in time. Our first subject is to understand how electrical transients are displayed in a digital circuit, specifically, how they are similar to and how they are different from the mechanical and quantum-mechanical phenomena in physical space. Both phenomena are a continuous sequence of events that develop in their respective space and time. The role of time remains the same, if we consider nonrelativistic problems. The cascaded chain of logic gates provides the space in which the switching transients are created and developed. Following the convention of physics, we consider the digital circuit a *field*, where the laws of circuit theory control creation, development and destruction of the electrical transient or excitation. In this chapter, I summarize the properties of the field of digital excitation, stressing the similarities to and differences from the phenomena in physical space. A gate-field is not a pure classical or a pure quantum-mechanical object. Therefore, it is hard to separate the discussion cleanly into the two limits. In Chapter 1, I discuss mostly the classical aspects of the gate-field. If I state my conclusion first, I am able to show that a multi dimensional cascaded mesh of digital gates has more similarities to than differences from physical space, where the activities of nature take place. The minor differences between the two add to the varieties of the electrical phenomena supported by the gate-field and make them quite attractive, once they are viewed by engineers. An equivalent circuit is able to model a real object, as well as an object that exists only in imagination. Electronic circuit theory as the science of models covers at least the entire physics and maybe more.

The most prominent similarity is the capability of propagating an excitation. An *excitation* is a spatially nonuniform profile of a field variable that moves with time. A particle in physical space is an excitation. The velocity of excitation can be measured, and it is its important attribute. Information is sent out from the source of the excitation when it is launched to the gate-field. The excitation carries information and energy. Two excitations interact. If two excitations meet at a location, we may say that a collision took place. Generally, it is impossible to distinguish two excitations when they overlap. At collision some information is lost. The energies carried by the two excitations can be converted to another energy form. The destination of the excitation consumes the energy. If we examine the extent of equivalence between the excitation and the digital signal, and between the physical space and the gate-field, the list of similarities is more impressive than the list of differences. This observation shows that a simple digital-gate field is able to display most, if not all, of the phenomena that occur in classical and quantum

Classical Gate Field

physical space.

Equipped with the essential capability of physical space to display classical and quantum phenomena, the structural simplicity, the variety, and the flexibility of the designing structure of a digital-gate field make a gate-field an interesting object of study. I may even hope to gain some insight into the mystery of the physical space itself, from the *tangible* gate-field model. Several attempts have been made in this book. To begin with, let us study the simplest example of a digital-gate field made from cascaded inverters. In the following study, the integrated-circuit technology used to build the gate-field does not matter essentially. We use the simplest and cleanest example of the complementary MOS technology. A number of excellent textbooks have been written on complementary metal-oxide-semiconductor (CMOS) digital-circuit technology, and readers are referred to them for background information (Kang and Leblebic, 1996; Glasser and Dopperpuhl, 1985).

1.02 Examples of the Gate Field

Figure 1.02.1(a) shows a single stage of a CMOS inverting buffer. The inverting buffer is powered by a pair of power-supply voltage levels, V_{DD} and V_{SS}. V_{SS} is often set at 0 and is called ground, and V_{DD} is often set at some positive voltage. The capacitance loading the output of the inverting buffer includes the input capacitance of the load stage, which cannot be set at zero because the field-effect transistor (FET) gate capacitance is essential for the FET operation (Shoji, 1992). The node capacitance is necessary if the buffer output node is physically observable. We note that a voltmeter is equivalent to a capacitance.

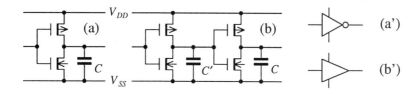

Figure 1.02.1 Inverting and noninverting buffers

This means that if the buffer's operation should be physically observable, the capacitance cannot be set at zero (Shoji, 1996). The inverting buffer generates the complementary output voltage in a steady state: if the input voltage to the buffer is $V_{SS} = 0$, the logic LOW voltage, the output voltage is V_{DD}, the logic HIGH voltage, and vice versa. If the input voltage is zero at the beginning and is increased slowly and continuously, the output voltage decreases from V_{DD} continuously. At a certain input voltage V_{GSW}, the increasing input voltage equals the decreasing output voltage. V_{GSW} is called the inverter-switching threshold voltage. A noninverting buffer is built by cascading two stages of the inverting buffers, as shown in Fig.1.02.1(b). Again, the capacitance loading at the output of each stage is required for the node to be observable. A buffer must be able to control the energy of the load capacitance. This means that at least one of the pullup or the pulldown FET must be on at any time. From this requirement, a FET source-follower circuit cannot be a noninverting buffer. At the HIGH or LOW steady state of the source follower both FETs are in the cutoff state, and they are unable to control the output capacitance energy.

In Fig.1.02.1(b), capacitances C' and C load the two nodes of the cascaded two stages of inverters that make a single noninverting buffer. Capacitance C is necessary, since the output-node voltage of the buffer should be observable. As for C' it depends on the modeling. If the internal node is observable, C' must be nonzero. If not, C' can be set at zero. Such a simplified noninverting buffer is used for certain special modeling. A question may arise as to how real such a noninverting buffer is, since it is only an idealization. This question is answered by examples in later sections of the book. The issue is related to the observability of the gate-field: certain parameter can be *defined* as unobservable at the time when the gate-field is defined. Here we assume that two types of buffers are available. Among the two types of the buffers, an inverting buffer is physically more real and fundamental than a noninverting buffer.

The buffers are cascaded in a long chain as shown in Figs.1.02.2(a) (inverting buffer chain) and (b) (noninverting buffer chain). Suppose that the left-end terminal voltage of the cascaded chain V_1 was set at V_{SS}, the LOW logic or Boolean-level a long time before.

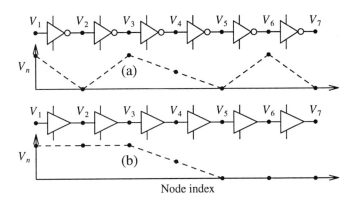

Figure 1.02.2 A one-dimensional digital buffer field

At time $t = 0$, the voltage makes a stepwise increase to V_{DD}, the HIGH logic level. The voltage profile of the chain sometime later is shown in

Classical Gate Field

the figure. In the inverting buffer chain the node voltages at the successive nodes alternate V_{DD} and V_{SS}, except at the location where the progressing wavefront exists. At the wavefront, the phase of the alternating HIGH-LOW sequence shifts by a 180-degree phase angle. In a noninverting gate chain the voltage profile is a progressing stepfunction voltage front that is simpler than the progressing alternating-voltage profile of the inverting buffer chain. Either wavefront moves to the right with time. The velocity of the wavefront is given by

$$\text{Velocity} = 1/T_D \quad \text{where} \quad T_D = fCR_{EQ}$$

where T_D is the buffer delay time, C is the capacitance loading the output node of the buffer, R_{EQ} is the resistance looking into the buffer from the output and f is a factor of the order of unity. The buffer delay is essentially the charge-discharge delay of the load capacitance through the inverter's output resistance. We note that the same formula holds in the noninverting buffer chain, since we set $C' = 0$ in Fig.1.02.1(b). The dimension of *velocity* is (Number of stages)/(Unit of time). Either buffer chain is able to support a moving digital excitation, very much as the physical space is able to support a traveling wave or a particle. The buffer chains possess the essential capability of one-dimensional physical space. We call such a structure a *gate field*.

A one-dimensional gate-field can be built from gates more complex than inverters and cascaded inverters as the building units. If a multiinput gate is used as the unit, the extra inputs of the gates are set at the logic level that enables the gate, but some of the inputs can be used to build a structure in the one-dimensional field. The gate location works as a launcher of digital excitation, a barrier to block further propagation of excitation, a controller of the excitation velocity, or a confluent point of two one-dimensional upstream gate-fields. There are many structural varieties.

A conventional logic gate is a multiple-input, single-output gate. A two-dimensional digital gate-field can be constructed from any of the conventional two-input logic gates like NOR2, OR2, NAND2, AND2, XNOR or XOR. At each point of the digital gate-field there is a gate, and it accepts signals from the two upstream gates. A two-dimensional AND gate-field is shown in Fig.1.02.3. At any point in a two-dimensional gate-field, the four nearests neighbor gates surround the gate at the location. Out of the four gates, the two upstream gates drive the gate, and the two downstream gates receive the output signal of the gate. In this way a two-dimensional signal net can be built. This is only one of the many connections of the gates, and there are many other connections, each having its own peculiar propagation characteristics.

In Fig.1.02.3, the gates are connected such that a signal propagates from left to right, and from bottom to top. $X_1, X_2, \ldots$, are the inputs. If X_n is permanently in the disabling LOW level, all the AND gates above it are disabled. The signal propagation stops at the boundary, and only the lower part of the field is able to support the activities.

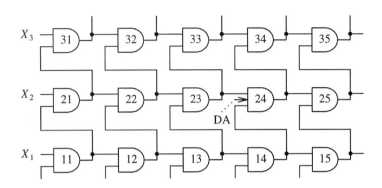

Figure 1.02.3 A two-dimensional AND gate-field

As in the one-dimensional gate-field, structures of a two-dimensional gate-field can be built by selectively enabling or disabling the multiinput gates. If gate 24 is converted to a three-input AND gate, and if the input DA is used to disable the gate, the region of the field at the upper right quadrant of gate 24 is disabled. In a gate-field assembled from only one type of logic gate, of the AND or OR type, a disable signal has a strong effect. One disable signal propagates downstream and keeps a large fraction of the field from the activities. There are many other ways to introduce nonuniform features into the gate-field. If three input gates are used to build the structure, the signal may propagate both above and below. Thus, a gate-field is a model of a complex connected web of digital gates in a logic system, as well as a restricted model of a physical space.

A gate-field that has a local, rather than a global multidimensional structure has many interesting characteristics. The structure of Fig.1.02.4(a) has a loop, called a *recombinant fanout*, at the input of the NAND gate, and the structure of Fig.1.02.4(b) has a loop that includes the NAND gate from the downstream back to the upstream. This structure may work as a local generator of periodic excitation. Either way, the structure reflects the properties of the local loop on the phenomena in the entire gate-field, such as creation or destruction of excitation. This is a counterpart of the Karuza's field model (Freedman and van Nieuwenhuizen, 1985).

The structure of a gate-field has rich varieties, since the connectivity between locations and the unit structure contained at a single location has great varieties. As for complexity at a single location, the circuit assigned to a location may have any degree of complexity.

Classical Gate Field

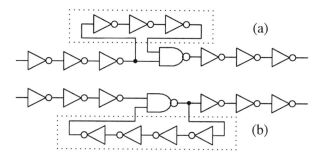

Figure 1.02.4 An extra dimension of a gate-field spun at a location

Depending on the circuit, some unusual propagation properties can be created: two examples are shown in Fig.1.02.5. In this figure, the bottom two examples of the circuits fit into the top array of boxes to make a one-dimensional gate-field. A gate-field consisting of simple logic gates does not support stable indefinite propagation of an isolated return-to-zero pulse (RZ pulse). The width of a narrow pulse decreases as it propagates, and finally the pulse collapses (Sections 2.17 and 2.18). The circuit in Fig.1.02.5(a) is a single-shot circuit that generates a pulse whose width is determined by the resistance-capacitance (RC) time constant of the circuit: every time the input is triggered by an upgoing pulse, a fixed-width pulse is generated at the output. If the circuit makes a gate-field, an isolated pulse propagates indefinitely along the field. The pulse width is maintained by the time constant at each location of the circuit, and it never diminishes. The gate-field built from the circuit of Fig.1.02.5(b) is a binary countdown chain. If the left end of the field is driven by a pulse train having period τ, the pulse period in the downstream increases by a factor of two for each stage, to $2\tau, 4\tau, \ldots$. The activities of the gate-field decrease, as the pulse propagates the downstream.

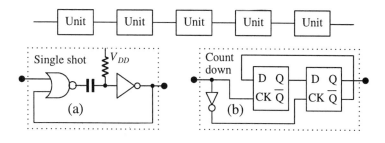

Figure 1.02.5 Examples of a complex gate-field

1.03 The Vector Gate Field

In the one-dimensional gate-field discussed in the last section, only one information-carrying signal moves from one location to the next. The number of signals can be increased, or a gate-field node location may have more than one distinguishable node. In this generalized structure, a qualitatively new requirement emerges: a composite signal must propagate synchronously as a single vector or, generally, as a tensor object. If not synchronized, a vector breaks down into its components as it propagates through the gate-field and that is equivalent to having so many independent and parallel gate-fields. A synchronized circuit of this type is used in conventional logic circuits. A cascaded programmable logic array (PLA) logic is an example. The physical space has the capability of propagating a vectorized signal, such as an electromagnetic field. The cascaded PLA circuit is a gate-field counterpart of the physical space that carries an electromagnetic wave.

As a simple example, the structure of a gate-field built by cascading tristatable inverters is shown in Fig.1.03.1. The circuit diagram of a tristatable inverter is shown to the right margin of the figure. If CK is HIGH and $\overline{CK}$ is LOW, a tristatable inverter is logically equivalent to a conventional CMOS inverter. If CK is LOW and $\overline{CK}$ is HIGH, however, the output node of the tristatable inverter is isolated from the rest of the circuit, and the output node voltage is sustained by the charge deposited previously in the node capacitor.

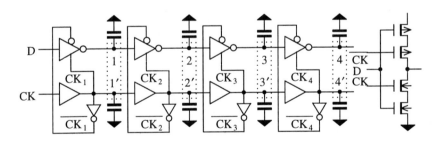

Figure 1.03.1 Cascaded tristatable inverters as a vector-gate field

In the circuit, both the tristatable inverters and the noninverting clock buffers have delays. The tristatable inverter chain nodes 1, 2, 3, 4, ..., and the clock buffer nodes 1', 2', 3', 4', ···, have node capacitances, and, accordingly, the node voltages are measurable. The pair of the node voltages at the same node index make a vector, but if quantized to make a set of Boolean-levels, they make a Boolean vector.

Let CK be HIGH, $\overline{CK}$ be low, and input node D be low. After

Classical Gate Field

some time all the even-numbered nodes of the tristatable inverter chain settle at the LOW logic level, and all the odd-numbered nodes settle at the HIGH logic level. Then CK makes a HIGH to LOW transition. All the node capacitances of the tristatable inverter chain are isolated from the rest of the circuit, while maintaining the same voltage levels that existed before the clock transition. This is the initial state of the tristatable inverter field. After the initial condition is established, nodes D and CK, which are the components of a vector at the location, both make the LOW to HIGH transition. As the first-stage tristatable inverter is released from the tristate, node 1 makes a HIGH to LOW transition. After the transition is over, the delayed clock signal begins to enable the second-stage tristatable inverter, and node 2 begins to pull up. Thus, the signal and the clock propagate together. Synchronization between the signal and the clock is maintained only if the delay of the tristatable inverter is significantly shorter than the delay of the clock buffer. Practical cascaded PLA logic circuits are designed to satisfy this condition. If the clock delay is shorter than the tristatable inverter delay, the clock propagates faster than the data, and the vector-propagation feature is destroyed. Obviously, there are number of variations of the vector variable propagation scheme. To maintain the integrity of a vector requires synchronization of multiple signals, and this is a rather strong restriction for the gate-field. Using a clock for this purpose is the most obvious method.

A variation of a vector-gate field is a bidirectional field. Since a logic gate has signal directionality associated with gain, a bidirectional propagation of excitation through a single path in a scalar field is not possible. If a bidirectional connection at each node of a noninverting scalar field is made, it applies a positive feedback through the buffer loop, and the circuit at each node becomes a latch that is stuck at one of the two possible states, thereby preventing signal propagation in either direction. If a bidirectional connection is made at each node of an inverting scalar gate-field, each location generally becomes an oscillator. Thus, bidirectional propagation is impossible by a scalar field. Suppose that a pair of gate-fields have opposite direction of propagation make a single composite field as shown in Fig.1.03.2. One location consists of a pair of nodes, X and Y. Both fields present a default LOW level. When an excitation is launched, both nodes are driven to the HIGH level. When a node is measured, it is determined to be HIGH if at least one node is HIGH. This forcing or measuring protocol may sound peculiar, but it is not. In a gate-field model, not everything that exists in the circuit is accessible to the observer as in a classical object, as we have seen in Fig.1.02.1, where node capacitance C' can be defined as unobservable. In this case, the only peculiarity is that the two nodes at the same location are not distinguishable or are made to be the same node by definition but are electrically separate. With this definition, the pair of gate-fields make a bidirectional field. In the physical space bidirectional transmission of electromagnetic wave is possible because the excitations allow linear superposition. In the nonlinear gate-field, two paths are required for two directions. Highway automobile traffic is an example. Since it is impossible to define what a *superposed* vehicle is (except possibly by the police), bidirectional traffic requires two independent lanes. A location on the highway is physically two points on two lanes, but on a highway map we call them by a single name. As such, a two-lane or wider highway is a vector field. A multiple-gate field having structure has wide flexibility. For instance, the amplitude and the propagation velocity to the right and to the left need not be the same. A reflective end of a bidirectional vector field couples the incoming signal either as it is or inverted to the output. In the former, the wave is reflected as it is, like an open end of an LC transmission line. In the latter, it is like a shorted end of the inductance-capacitance (LC) line. An absorbing end of the field has the incoming end disconnected and the outgoing end grounded.

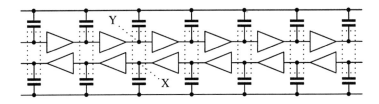

Figure 1.03.2 Bidirectional propagation of excitation

The combination of transition edges can be looked at from a different angle. In a gate-field a pair of downgoing and upgoing transition edges make a pair of particle and antiparticle like complementary objects. To make a traveling pair of a particle and an antiparticle, a gate-field has two parallel lines that carry the complementary pair of signals. If the pair of nodes is observed by a coarse observer who is unable to recognize the individual nodes, the vector carries no information. If the observer is able to recognize the node activities separately, the bidirectional transmission will be recognized. The distinguishability of nodes is, again, a matter of definition. Then a question arises: what is really observable? It is energy, but it has a complex of meanings in a gate-field, as is shown in the next section.

1.04 Energy Transfer in Gate Field

Energy and matter had been the two clearly distinct objects of physics until the nineteenth century. With the advent of the theory of relativity, a new viewpoint that matter is a representation of energy emerged. Energy is more fundamental than matter, since even in classical physics, measurement of any physical variable requires energy. Nonexistence of energy makes physical observation impossible, and therefore physics loses its supporting evidence. Many concepts of modern physics especially the concepts of solid-state physics, such as photons, polarons, magnons, and solitons are based on the new viewpoint that identifies energy as a visible and tangible object. They are many examples of the equivalence and duality of energy and matter, or particles. By extending this line of thought, digital excitation in a logic circuit, which is a local concentration profile of energy, may be considered as matter, or more descriptively, as a particle that follows the laws of mechanics. This book is based on this viewpoint. Because of the many unique features of the gate-field, many details of this idea must be explored.

Each node of a gate-field is a node of the logic circuit loaded by a capacitance to ground, C. This is the fundamental requirement of observability (Shoji, 1996). If the node voltage is V, the capacitance holds energy $E = (C/2)V^2$. This is the energy associated with the node. The energy is supplied from the power source V_{DD} and V_{SS} (or ground). With reference to Fig.1.04.1(a) if the input of the inverter is pulled down from the HIGH to LOW logic level, the N-channel FET (NFET) turns off, and the P-channel FET (PFET) turns on. Current flows from V_{DD} through the channel of the PFET to capacitor C and to the ground. The direction of the current is shown by the solid arrow segments. A positive charge flows into the upper plate, and a negative charge flows into the lower plate of the capacitor symbol. When C is charged to V_{DD}, it acquires the maximum energy available from the power source $(C/2)V_{DD}^2$. If C and V_{DD} are constants over the field, the energy is a constant of the field as well. The energy delivered by the power supply to the circuit is *not* equal to the energy that the capacitor has received. Looking from the power supply, charge $Q = CV_{DD}$ moved from potential V_{DD} to potential $V_{SS} = 0$. This may not be clear, since C retains charge CV_{DD} after the charging process. But as far as it is observed from the power supply, the charge in C never matters: charge Q was sent out from the V_{DD} and was received by the V_{SS}. If the plus and minus charges of the capacitor plates are neutralized, it is clear that charge Q indeed moved from V_{DD} to V_{SS}. Instead of complete neutralization by one step (the plus and the minus charges mixed completely), the charges $\pm Q$ were brought into an immediate proximity of the capacitor plates, and that retains the residual energy $(C/2)V_{DD}^2$, which is half the energy supplied by the power source. The energy loss of the power supply, $Q \cdot V_{DD} = CV_{DD}^2$, is twice the energy retained by C. The balance of the energy was converted to heat by the resistance of the channel of the charging PFET. This is the scenario if the input of the inverter switched within zero time. If it takes nonzero time to effect the transition, some additional energy is lost by the current that flows through the channels of the PFET to that of the NFET connected in series. Then $Q > CV_{DD}$, and the energy lost by the power supply QV_{DD} is higher than CV_{DD}^2. If the input of the inverter is subsequently pulled up within zero time as shown in Fig.1.04.1(b), the capacitor charge CV_{DD} is neutralized by the current of the NFET channel, and the energy $(C/2)V_{DD}^2$ is dissipated as heat generated in the NFET channel. During the neutralization process the power supply is isolated by the nonconducting PFET, and no energy exchange between the power supply and the inverter circuit as a whole takes place.

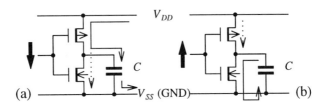

Figure 1.04.1 The charge-discharge process of node capacitance

Suppose that there is an observer who has a voltmeter that is used to measure the logic-node voltages. While the switching proceeds and the energy exchange takes place, the observer of the gate-field recognizes only the voltages of the nodes of the gate-field or, equivalently, the energies of the capacitors at the gate-field nodes. They are the only gate-field variables that are observable. Specifically, the currents of the FETs are not observable. Suppose that input at the left end of the inverter field of Fig.1.04.2 was set at a LOW logic level a long time before and that then the input voltage makes a stepwise increase to a HIGH level at time $t = 0$. What will an observer of the gate-field see? The observer has no knowledge of how the logic-gate circuit works together with the power supply. They are all hidden behind the scenes. He observes that the capacitor at location A loses energy and that the capacitor at location B gains energy. As the voltage wave passes locations A and B, the observer recognizes that energy at A moved to B. In reality the energy at location A was lost as heat, and then energy was supplied to location B via the power source by the mechanisms shown in Fig.1.04.1, but the observer has no way of knowing the behind-the-scenes mechanisms.

This may sound like a strangely restricted description of the gate-field, but that is all that the observer of the logic circuit or a technician is able to access. In practice, large-scale logic-system troubleshooting has always been carried out in this way. If the observer has a logic probe instead of a voltmeter, he knows even less only a digitized reading of the node voltage. Is there anything unreasonable? No. When we examine an elementary particle created in a physical vacuum, do we really know everything about the structure of the background vacuum that created the particle? No. The digital system technician and the elementary particle physicist have a lot in common that is, lack of knowledge of the

Classical Gate Field

background. This is the basic feature of the gate-field model.

Let us consider this problem more closely. The following criticism may be raised: (1) If only nodes C and D of Fig.1.04.2 are observed, node C acquires energy before node D loses it; (2) After the transient passes the two locations, it looks as if energy moved backward from D to C. Backward transfer of energy appears inconsistent with the forward-propagation direction, but this is not really a contradiction. How does one interpret this observation of an unfortunate choice of a pair of nodes? Apparent backward transfer of energy does not contradict forward propagation of the front. As a familiar example, forward motion of a bubble in a capillary is really a backward motion of water. This confusing conclusion could never have been reached if more than two nodes were observed simultaneously while tracking the development of the transient. Only the logic nodes are observable, but they are all simultaneously observable. Then the direction of the movement of the front and of energy can be unmistakably established. Yet this is a good example to show that by restricting access to the object we may reach all sorts of qualitatively different conclusions from the experiments. An interesting point is that restricting access to the object *properly* results in the conclusions that are mutually consistent. This is equivalent to the *hidden variable* interpretation of quantum mechanics, generally not accepted as correct (Bohm, 1951). What is interesting to us is that if the model of a system having hidden variables is created, the system may work very much like a quantum-mechanical object. This is the gate-field model.

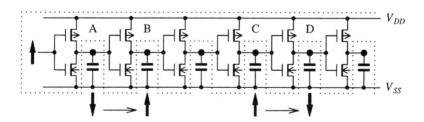

Figure 1.04.2 Visible energy transfer in a gate chain
and apparent energy borrowing

The direction of motion of the excitation can be unmistakably identified by observing many nodes simultaneously, but the energy balance of nodes C and D must be interpreted. In a gate-field the energy is conserved only when the steady state is reached. Within the period of time comparable to the gate's switching delay time, if node C can borrow energy from the hidden background of the gate-field, and if node D pulls down, the borrowed energy appears to be returned to the hidden background. In this view, the energy conserves only if the states are steady states. This interpretation makes the gate-field quite similar to quantum-mechanical physical space, where energy-time complementarity holds.

Is this an outrageous explanation? I do not think so. Since the advent of quantum field theory, the quantum-mechanical vacuum has never been characterized as a simple emptiness having no structure. The mechanism that supports excitation such as an elementary particle is quite complex, and the observer has only limited access to it. As was speculated, not even the dimension of a physical vacuum in small spaces and time scales is certain (Freedman and van Nieuwenhuizen, 1985). In both cases we correlate only what is observable, and a hidden background should not take part in interpretation. In doing so we must come up with the most rational and consistent interpretation.

The scale of the apparent movement of energy in a gate-field depends on the type of gate. In an inverting buffer field, shown in Fig.1.04.3(a), the observed movement of energy is from one node to its downstream node. Since the driving node is considered to be energized by the signal source, energy moves as is indicated by the arrows. No large-scale energy transfer across the field is observed. In the noninverting gate-field of Fig.1.04.3(b) the capacitors at the nodes are charged sequentially from the left end by the signal source, as the excitation propagates. The observer sees energy being supplied by the signal source and being transferred across the entire gate-field. In Fig.1.04.3(a) the left end of the inverter chain was set at LOW logic level and was driven up to HIGH logic level. If the transition polarity is inverted, two alternative explanations are possible. Either the direction of energy transfer switches (from node N to node $N-1$), or the node identification is changed, such that energy is transferred from node $N-1$ to node N instead of from N to $N+1$, and this time the signal source receives energy from the field. The second interpretation is rational because it shows energy exchange between the field and the source as the input switches, but the first interpretation has merit because the energy-flow direction simply reversed. We need to contend, if the observation is restricted, that certain basic issues cannot be decided unambiguously, since there is no way to tag energy. This is the basic nature of an elementary object. It does not matter whether the elementarity is the nature of the object itself or whether it is the consequence of the unavailability of information. The gate-field has two stable states: in the one the odd-numbered nodes are HIGH, and even-numbered nodes are LOW; and vice versa in the other. The excitation propagation is the means of switching the state of the gate-field from the one to the other. How much the switching has progressed can be determined by locating the excitation front, but the sort of energy transfer that effected the switching cannot be decided beyond a certain limit. In the noninverting gate-field of Fig.1.04.3(b) the interpretation is easier: as the input transition polarity

Classical Gate Field

reverses, the direction of the energy flow across the gate-field reverses direction as well, from left to right to right to left, including the signal source. The point is that we really should not put too much emphasis on the apparent contradiction.

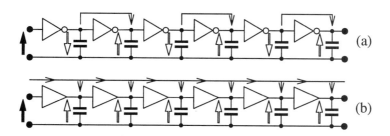

Figure 1.04.3 Apparent movement of energy in a gate-field

Let us observe how charge and energy are apparently transferred from one node to the next under the restricted observation conditions. In the CMOS inverter chain of Fig.1.04.2, the FETs are modeled by the low-field, gradual-channel model (Brews, 1981). NFET and PFET currents, I_N and I_P, respectively, are given by the model as

$$I_N = B_N[V_{n-1} - V_{THN} - (1/2)V_N]V_N$$
$$(V_{n-1} \geq V_{THN} \text{ and } V_n \leq V_{n-1} - V_{THN})$$
$$I_N = (B_N/2)(V_{n-1} - V_{THN})^2$$
$$(V_{n-1} \geq V_{THN} \text{ and } V_n \geq V_{n-1} - V_{THN})$$
$$I_N = 0 \quad \text{(otherwise)}$$
$$I_P = B_P[V_{DD} - V_{n-1} - V_{THP} - (1/2)(V_{DD} - V_n)](V_{DD} - V_n)$$
$$(V_{n-1} \leq V_{DD} - V_{THP} \text{ and } V_n \geq V_{n-1} + V_{THP})$$
$$I_P = (B_P/2)(V_{DD} - V_{n-1} - V_{THP})^2$$
$$(V_{n-1} \leq V_{DD} - V_{THP} \text{ and } V_n \leq V_{n-1} + V_{THP})$$
$$I_P = 0 \quad \text{(otherwise)}$$

where V_{n-1} and V_n are the $(n-1)$-th and n-th node voltages, respectively. The circuit equations satisfied by V_ns are

$$C(dV_n/dt) = I_P - I_N \quad (0 < n \leq n_{max})$$

where n_{max} is the length of the gate-field used in the numerical analysis. The set of equations are solved numerically, using the parameter values

$$n_{max} = 20 \quad n = 16 \quad B_N = B_P = B = 1.0 \quad C = 1.0$$

$$V_{THN} = V_{THP} = 0.1 \quad V_{DD} = 1.0$$

and subject to the initial condition that the odd-numbered nodes are originally CMOS HIGH and the even-numbered nodes are originally LOW logic levels. Charge Q_n and energy E_n of node $n = 15$ and 16 are defined by

$$Q_n = CV_n \quad E_n = (C/2)V_n^2$$

and are plotted versus time, as shown in Figs 1.04.4(a) and 1.04.4(b), respectively.

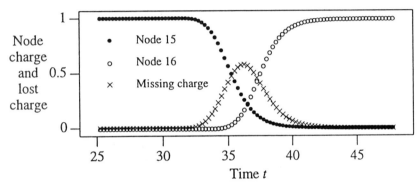

Figure 1.04.4(a) Time dependence of node charge (20-member chain, from node 15 to 16)

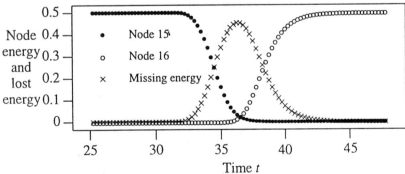

Figure 1.04.4(b) Time dependence of node energy (20-member chain, from node 15 to 16)

As node 15 loses charge or energy, node 16 gains it. The × signs in Figs.1.04.4(a) and (b) are, respectively, the charge and the energy that are observed as temporarily lost, defined by

$$\Delta Q(t) = CV_{DD} - Q_{15}(t) - Q_{16}(t) \quad \text{and}$$
$$\Delta E(t) = (C/2)V_{DD}^2 - E_{15}(t) - E_{16}(t)$$

During the switching process the observed energy does not conserve: the energy appears lost in the middle of the transient. The energy is eventually recovered as the transient completes. Energy and charge conservation is violated for the restricted observer during the transient. This may sound irrational, but in a system that has a background structure like the gate-field, this is not unusual, if we look at the entire system. Metaphorically, the observer and the object are not like a physicist and a classical mass point but rather like an actor on a stage and a spectator in a theater. In the following I discuss a detailed mechanism for signal transmission through an LC transmission line that shows, similarly, an apparent loss of energy in the process of transmission, if the observation is restricted.

From this viewpoint, I cannot help boldly speculating the following. Energy non-conservation, or the energy-time uncertainty principle of quantum mechanics,

$$\Delta E \cdot \Delta t > h$$

at the time of creation and annihilation of a particle may suggest a behind-the-scenes energy exchange since the physical vacuum posesses a complex structure. I dare to propose that a deeper study of the phenomena in a gate-field might provide some new insights into the issue. Indeed, model building is much easier in circuit theory than in physics, and that is a fascinating aspect of modern circuit theory. Which other branch of physics offer such convenience?

Let us estimate the value of h, which controls the uncertainty. Since ΔE and Δt are estimated at

$$\Delta E \approx (C/2) V_{DD}^2 \qquad \Delta t \approx C/BV_{DD}$$

we have

$$h \approx \frac{CV_{DD}^2}{2} \cdot \frac{C}{BV_{DD}} = \frac{C^2 V_{DD}}{2B} \quad (erg \cdot sec)$$

This is the parameter that plays the role of the Planck's constant of the gate-field.

One of the enduring mysteries of quantum mechanics is related to the fine-structure constant f_Q given by

$$f_Q = q^2/hc = 1/137. \cdots$$

where q is the electronic charge, h is the Planck's constant, and c is the light velocity. This is a nondimensional number including the fundamental constants of physics only. Why the fundamental constants are related by the simple relationship, and why the number is the inverse of 137.... has never been undisputably explained, and this is considered to be one of the hard problems of physics (Heitler, 1945). Since we are making a gate-field model very similar to the quantum-mechanical field, again we cannot avoid the temptation to inquire about the corresponding parameter of the gate-field. We derived the equivalent of the Planck's constant, and the velocity of propagation is derived in Section 1.06

$$h = \frac{C^2 V_{DD}}{2B} \qquad v \approx \frac{BV_{DD}}{C}$$

Since it is impossible to propagate an excitation in a gate-field faster than v, we may identify $c = v$. To correlate f_Q with the parameters, we need to write

$$f_Q = \frac{(q^2/a)}{h} \cdot \frac{a}{c}$$

where a is the size of the excitation. By doing so we introduce a special interpretation into the problem: the interpretation is only one of the possible alternative interpretations in *dimensional* analysis. The expression q^2/a is the self-energy of the excitation, and a/c is the time required for an excitation to travel a distance equal to its dimension, or the delay time of the gate. The parameters are substituted as

$$\frac{q^2}{a} \rightarrow \frac{C}{2} V_{DD}^2 \qquad \frac{a}{c} \rightarrow \frac{C}{BV_{DD}}$$

By substituting them, f_G, the corresponding parameter of the gate-field is

$$f_G \approx \frac{CV_{DD}^2}{2} \cdot \frac{2B}{C^2 V_{DD}} \cdot \frac{C}{BV_{DD}} = 1$$

The gate-field equivalent of the fine-structure constant is about unity. There are two orders of magnitude difference between f_Q and f_G. Explanation of the difference may provide some insight into the nature of the mysterious parameter. This argument is perhaps too crude to extend beyond this point, but the fundamental idea that a concrete and real model of the physical space is necessary to explain the number will be undisputable.

Violation of the observed-energy conservation law during the process of energy transfer from a node capacitor to a node capacitor in the excitation transmission path is a commonly observable phenomenon that exists even in a conventional LC transmission line. The LC transmission line of Fig.1.04.5(a) is a close parallel to a gate-field, except that the line has no apparent behind-the-scenes energy-supply mechanism or the power supply. The energy conservation of the excitation that propagates along the LC line is clear, since both L and C are lossless. In this model our observational restriction is crucial that is, only the voltages developed across the capacitors are observable, but the currents in the inductances are not observable. This is why the missing energy interpretation becomes necessary. As it was discussed in my last book, the voltage of an inductive circuit must be measured by a small voltmeter set up at the geometrical center of the capacitance plates that closes a loop to make the loop inductance (Shoji, 1996). The observable part of the LC transmission line field is enclosed in the dotted lines A. Inductance L plays the role of a gate in a gate-field, and its current and magnetic energy are not accessible to the observer who has only a voltmeter. Indeed, we may say that there is no way to measure the magnetic energy of the inductance by a conventional logic probe. In this circuit model there is one more restriction discussed in my last work (Shoji, 1996). Since the inductance is the property of a loop, it does not make sense to measure the voltage developed across the two nodes of the inductance, as it is shown in the equivalent circuit diagram. Including the inductance in the background, the entire system conserves energy all the time, but that is not what the restricted observer of the capacitor voltages sees. From the capacitor energy observation alone, the magnetic energy of the inductance appears lost. Then the energy conservation is valid only if two steady states are compared. Let us consider how the energy of capacitor C_1 of a long LC transmission line is transferred to the next location C_0.

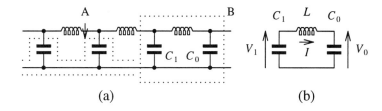

Figure 1.04.5 Energy transfer mechanism in an LC chain

To simplify the problem we consider only one section of the LC chain circuit enclosed by the dotted rectangle B of Fig.1.04.5(a), which is shown in Fig.1.04.5(b). The line is charged from the left end, and it has been charged up to C_1. The problem is how C_0 is charged by C_1. C_1 represents the entire charged section to the left side, so we chose $C_1 \gg C_0$. We define voltages V_1, V_0, and current I as shown in Fig.1.04.5(b). At $t = 0$ we have

$$V_1(0) = V_{DD} \quad V_0(0) = 0 \quad I(0) = 0$$

and they satisfy the circuit equations

$$C_0[dV_0(t)/dt] = -C_1[dV_1(t)/dt] = I(t)$$
$$L[dI(t)/dt] = V_1(t) - V_0(t)$$

The equations are solved numerically, using the following parameter values:

$$C_1 = 5.0 \quad C_0 = 1.0 \quad L = 1.0 \quad V_{DD} = 1.0$$

The energy of nodes 1 and 0 are defined by
$$E_1(t) = (C_1/2)V_1(t)^2 \quad E_0(t) = (C_0/2)V_0(t)^2$$
and the apparent loss of energy
$$\Delta E(t) = E_1(0) - E_1(t) - E_0(t)$$
is determined. The results are shown in Fig.1.04.6. The energy that is not observable from the node-voltage measurement is stored as the magnetic-field energy of the inductance outside the electrically connected part of the circuit or in the core region of the inductance. Since the test equipment is insensitive to magnetic field, the magnetic energy appears lost. Temporal loss of energy is again a common phenomenon in a system that allows signal propagation and is accessible only by limited observations. The energy conservation holds in the steady state even from limited observation, but it does not hold in the transient state, to the degree determined by the structure behind the scene.

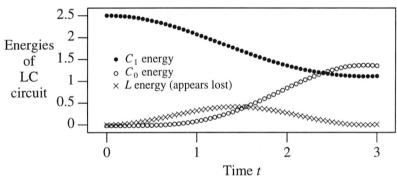

Figure 1.04.6 Energy transfer in an LC circuit

1.05 CMOS Inverter Switching Process

To determine the excitation velocity in a uniform gate-field, two issues must be resolved: (1) How to determine the time of arrival of an excitation at a location (inverse of the delay time, the difference between the arrival times of the excitation at the output and at the input is the velocity of the excitation) and (2) how to determine the waveform of the excitation. Since the first problem is a fundamental issue related to the classical and quantum-mechanical model, its discussion is postponed to Chapter 2. Here we use the conventional (classical) method to determine it at a chosen logic-threshold voltage. We concentrate on resolving the second issue. To do so, we need to determine a CMOS inverter switching waveform that is driven by an arbitrary input waveform. This problem is formulated mathematically as follows. Let the power supply, the input voltage waveform, and the output voltage waveform be V_{DD}, $V_1(t)$, and $V_0(t)$, respectively. The PFET pullup current and the NFET

Classical Gate Field

pulldown current are I_P and I_N, respectively, and the load capacitance is C, as shown in Fig.1.05.1(a). The circuit equation is then written as

$$C \frac{dV_0(t)}{dt} = \Delta I[V_{DD}, V_1(t), V_0(t)] \qquad (1.05.1)$$

where $\Delta I[...] = I_P[V_{DD}, V_1(t), V_0(t)] - I_N[V_1(t), V_0(t)]$ is the current that charges or discharges the load capacitance C, depending on its sign. I_P and I_N are the functions of the PFET and the NFET bias points, respectively, determined by $V_1(t)$ and $V_0(t)$. Time depencence of $V_1(t)$ and $V_0(t)$ is explicitly written for convenience. In a conventional inverter-delay analysis, $V_1(t)$ is given as a known function of time t, and $V_0(t)$ is determined by solving the differential equation. This is conventional problem solving, and its results are deterministic: for any $V_1(t)$, a $V_0(t)$ exists. Since $\Delta I[...]$ is a complex function of V_1 and V_0, the problem is mathematically quite difficult.

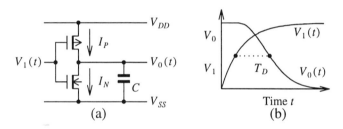

Figure 1.05.1 A CMOS inverter switching analysis

Let us observe the mathematical difficulty closely. I_P and I_N given by the gradual-channel, low-field FET model are explicitly written as follows:

$$I_P = 0 \quad (V_1 \geq V_{DD} - V_{THP}) \qquad (1.05.2)$$

$$I_P = B_P[(V_{DD} - V_1 - V_{THP}) - (1/2)(V_{DD} - V_0)](V_{DD} - V_0)$$

$$(V_1 < V_{DD} - V_{THP} \quad \text{and} \quad V_0 > V_1 + V_{THP})$$

$$I_P = (B_P/2)(V_{DD} - V_1 - V_{THP})^2$$

$$(V_1 < V_{DD} - V_{THP} \quad \text{and} \quad V_0 \leq V_1 + V_{THP})$$

$$I_N = 0 \quad (V_1 \leq V_{THN}) \qquad (1.05.3)$$

$$I_N = B_N[(V_1 - V_{THN}) - (1/2)V_0]V_0$$

$$(V_1 > v_{THN} \quad \text{and} \quad V_0 < V_1 - V_{THN})$$

$$I_N = (B_N/2)(V_1 - V_{THN})^2$$

$$(V_1 > V_{THN} \quad \text{and} \quad V_0 \geq V_1 - V_{THN})$$

$\Delta I(V_{DD}, V_1, V_0)$ is a function of the two variables V_1 and V_0. The ranges of V_1 and V_0 in a conventional digital circuit operation condition are

$$0 \le V_1, V_0 \le V_{DD}$$

ΔI is then a function within the square two-dimensional region of a V_0 - V_1 coordinate plane. Figures 1.05.2(a) and (b) show plots of ΔI within the region, by many equal height contours, like a conventional topographic landmap. The parameter values used for are
Figure 1.05.2(a): $B_N = B_P = 1.0$, $V_{DD} = 1.0$, $V_{THN} = V_{THP} = 0.1$ (symmetrical inverter)
Figure 1.05.2(b): $B_N = 2.0$, $B_P = 0.5$, $V_{DD} = 1.0$, $V_{THN} = V_{THP} = 0.1$. (asymmetrical inverter)
and the contours A, B, C, D, ..., have heights (the values of ΔI) set as follows: A (+0.4), B (+0.3), C (+0.2), D (+0.1), E (0.0), F(-0.1), G (-0.2), H (-0.3), I (-0.4), J (-0.5), K (-0.6), L (-0.7), and M (-0.8). At the lower left corner where $V_1 = V_0 = 0.0$, ΔI has the maximum (positive), and at the diagonally opposite upper right corner where $V_1 = V_2 = V_{DD} = 1.0$, ΔI has the minimum (negative). The maximum and the minimum of Fig.1.05.2(a) are $B_P(V_{DD} - V_{THP})^2/2 = 0.405$ and $-B_N(V_{DD} - V_{THN})^2/2 = -0.405$, respectively. In Fig.1.05.2(b) the maximum and the minimum are +0.2025 and -0.81, respectively. Figure 1.05.2(a) is the ΔI of a pullup-pulldown symmetrical inverter, and Fig.1.05.2(b) is that of an asymmetrical inverter, whose pulldown NFET is four times larger than its pullup PFET. The intersection of curve $\Delta I(V_{DD}, V_1, V_0)$ and the V_1 - V_0 coordinate plane [or a trace $\Delta I(V_{DD}, V_1, V_0) = 0$ or the curve with identifier E] is a curve that represents the static input-output characteristic of the inverter. The dotted lines of Figs.1.05.2(a) and (b) divide the V_1 - V_0 domain into the three regions: above the two lines, the PFET is in the triode region, and the NFET is in the saturation region; Below the two lines, the NFET is in the triode region, and the PFET is in the saturation region; Between the dotted lines both FETs are in the saturation region. In the three regions ΔI depends on V_1 and V_0 differently. The ΔI is defined by different functions in the three different regions. Then ΔI is not an analytic function of the variables V_0 and V_1.

If Eqs.(1.05.2) and (1.05.3) are substituted into Eq.(1.05.1) and the resultant differential equation is inspected, the equation cannot be solved in a closed form. The fundamental mathematical difficulty is traced back to the FET characteristic, which is a nonlinear function of V_0. Then the differential equation cannot be solved by any known method. Except when the FET characteristic is quite simple, such as the collapsable current generator model (Shoji, 1992), Eq.(1.05.1) cannot be solved in a closed form. This mathematical difficulty prevented detailed analyses of CMOS inverter-switching processes. Very little is known about the processes that require an exact solution of the equation, such as the overlap current and the dependence of the inverter-switching delay time on the input-signal rise and fall times. These problems require a complete analysis of the inverter-switching process in a closed form. They have been studied by numerical simulation and not by a closed-form analysis, which always lacks general perspective. Interestingly, this mathematical problem can never be solved, but it can be circumvented. The solution of the redefined problem is found in a closed form, which gives a lot of insight into the CMOS inverter-switching process.

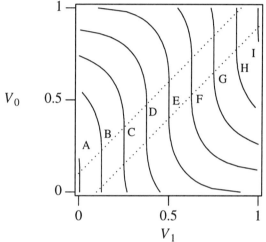

Figure 1.05.2(a) A contour map of ΔI

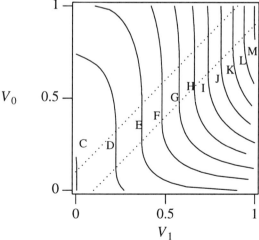

Figure 1.05.2(b) A contour map of ΔI

An accurate closed-form solution is not available if $V_1(t)$ is given and $V_0(t)$ is to be solved. If $V_0(t)$ is given, however, $V_1(t)$ can be solved by algebra only. From Eq.(1.05.1) $V_1(t)$ can be solved as a function of V_0 and $dV_0(t)/dt$ as

$$V_1(t) = \Phi[V_{DD}, V_0(t), dV_0(t)/dt]$$

and this relationship is of course rigorous: it is an accurate relationship between the input and the output waveforms. This method does not conform to the conventional physical-analysis method, however, and there are difficulties in thinking about the problem this way, since the mind is used to thinking from cause V_1 to effect V_0. Yet the solution method is interesting because certain peculiarities attributable to the reversal of the cause and effect relationship in problem solving are highlighted. Let us execute the mathematical procedure to the gradual-channel, low-field FET model of Eqs.(1.05.2) and (1.05.3) as a nontrivial example. The same procedure works for any FET model. We consider an inverter output pulldown transient. The algebra is straightforward but is tedious. Only the results are presented.

(1) If PFET is in the triode region and NFET is in the saturation region (an early phase of the output-pulldown transient), $V_{DD} - V_{THP} > V_1 > V_{THN}$ and $V_{DD} > V_0 > V_1 + V_{THP}$. We then have

$$V_1 = V_{THN} - (B_P/B_N)(V_{DD} - V_0) + \sqrt{X_1} \qquad (1.05.4)$$

where

$$X_1 = (B_P/B_N)^2(V_{DD} - V_0)^2 - 2V_{DD}[C/(B_N V_{DD})][dV_0(t)/dt]$$
$$+ (B_P/B_N)(V_{DD} - V_0)[V_{DD} - 2(V_{THN} + V_{THP}) + V_0]$$

(2A) If PFET and NFET are both in the saturation region (in the middle of the transient), $V_{DD} - V_{THP} > V_1 > V_{THN}$ and $V_1 + V_{THP} > V_0 > V_1 - V_{THN}$. We then have

$$V_1 = V_{THN} \qquad (1.05.5)$$
$$- [B_P/(B_N - B_P)](V_{DD} - V_{THP} - V_{THN}) + \sqrt{X_2}$$

where

$$X_2 = [B_P/(B_N - B_P)]^2 (V_{DD} - V_{THP} - V_{THN})^2$$
$$- 2V_{DD}[B_N/(B_N - B_P)][C/(B_N V_{DD})][dV_0(t)/dt]$$
$$+ [B_P/(B_N - B_P)](V_{DD} - V_{THP} - V_{THN})^2$$

(2B) In a special subcase, if $B_N = B_P$, the formula is reduced to

$$V_1 = V_{THN} + (1/2)(V_{DD} - V_{THP} - V_{THN}) \qquad (1.05.6)$$
$$- [V_{DD}/(V_{DD} - V_{THP} - V_{THN})][C/(B_N V_{DD})][dV_0(t)/dt]$$

(3) If PFET is in the saturation region and NFET is in the triode region (in the last phase of the transition), $V_{DD} - V_{THP} > V_1 > V_{THN}$

and $V_1 - V_{THN} > V_0 > 0$. We then have

$$V_1 = V_{DD} - V_{THP} + (B_N/B_P)V_0 - \sqrt{X_3} \qquad (1.05.7)$$

where

$$X_3 = [V_{DD} - V_{THN} - V_{THP} + (B_N/B_P)V_0]^2$$
$$+ 2(B_N/B_P)[C/(B_N V_{DD})]V_{DD}[dV_0(t)/dt]$$
$$- (V_{DD} - V_{THN} - V_{THP})^2 - (B_N/B_P)V_0^2$$

(3B) In a special case of (3) $V_1 > V_{DD} - V_{THP}$ and the PFET is turned off, and the NFET only controls the discharge process of the capacitance. In this regime V_1 must continue smoothly to the solution of regime (3). The solution of this case is given by

$$V_1 = V_{THN} + (1/2)V_0 \qquad (1.05.8)$$
$$- (V_{DD}/V_0)[C/(B_N V_{DD})][dV_0(t)/dt]$$

The case where both NFET and PFET are in the triode region never occurs.

Figures 1.05.3(a) and (b) show the V_1 solution as functions of V_0 and $dV_0(t)/dt$. The part of the range that belongs to region 2A or 2B is shown by the closed circles, and the right side is region 1 and the left side is region 3. The part above the dotted line $V_1 = V_{DD} - V_{THP}$ is region 3B. Each curve is for a fixed value of derivative (dV_0/dt), whose time scale is normalized using the time constant of the NFET, $C/(B_N V_{DD})$, as

Normalized derivative $= [C/(B_N V_{DD})][dV_0(t)/dt]$

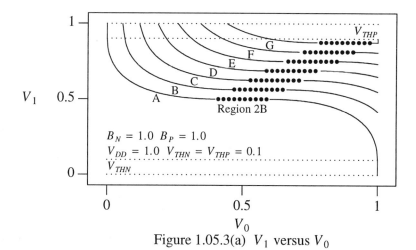

Figure 1.05.3(a) V_1 versus V_0

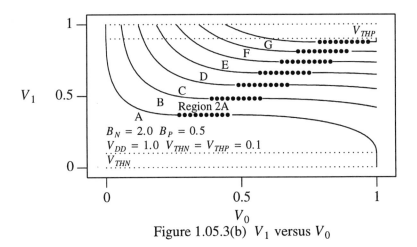

Figure 1.05.3(b) V_1 versus V_0

and the values used in the figures are zero or negative derivative values, since we consider an inverter pulldown transition. The symbols A, B, C, ..., in the figures indicate A(0.00), B(-0.05), C(-0.10), D(-0.15), E(-0.20), F(-0.25), and G(-0.30). Curve A (for zero derivative value) is the static input-output characteristic. Since V_0 is now the horizontal axis, the conventional I/O characteristic is reflected to the diagonal, $V_0 = V_1$. The constant values of V_1 of the curve A in the region 2A or 2B of Fig.1.05.3(a) and (b) are the switching threshold voltages of the inverter, V_{GSW}. In Fig.1.05.3(a) the inverter is symmetrical, and $V_{GSW} = 0.5$. In Fig.1.05.3(b) the inverter whose NFET is more conductive than the PFET, is asymmetrical. Then V_{GSW} is lower than 0.5 and is at about 0.4.

The inverter input waveform $V_1(t)$ can be determined from a given output $V_0(t)$ waveform by using the solution. Figure 1.05.4 shows several examples for different choices of $V_0(t)$. The choice of $V_0(t)$ becomes progressively realistic from Fig.1.05.4(a) to (c). In Fig.1.05.4(a),

$$V_0(t) = V_{DD} \ (t < 0) \ = V_{DD}[1-(t/T_B)] \ (0 < t < T_B)$$
$$= 0 \ (t > T_B)$$

The parameter values are given in the figure. The $V_1(t)$ waveform consists of the three regions 1, 2B, and 3 (note that $B_N = B_P$ in this example). The closed circles represent region 2B: to the left of that, region 1, and to the right of that, region 3 pulldown processes operate, respectively. $V_1(t)$ diverges at $t = T_B$, where the PFET turns off, and the pulldown transient that is still continuing thereafter is the work of the NFET only. The NFET carries the same current in spite of the reducing V_0, and this means $V_1 \to \infty$. This is an unrealistic operation of the circuit, and therefore the analysis must be terminated before $t = T_B$. The time required to complete the pulldown transient, $T_B = 20$, is long. Since

Classical Gate Field

the inverter switches gradually, or it operates quasi-statically, the switching mode is called a *static* switching mode.

In Figs. 1.05.4(b) and (c), $V_0(t)$ is given by a simple segmented formula

$$V_0(t) = V_{DD} - \alpha t^3 \quad (0 < t < T_A) \text{ that is} \quad (1.05.9a)$$

$$(V_{DD} > V_0(t) > V_{DD} - V_{THP})$$

$$V_0(t) = \quad (1.05.9b)$$

$$(V_{DD} - V_{THP}) \exp[-(t - T_A + \beta(t - T_A)^2)/t_0] \quad (t > T_A)$$

where the two segments of the V_0 curve join smoothly at $t = T_A$. At this time, we have $V_0 = V_{DD} - V_{THP}$. $V_0(t)$ from the two sides match in their values as well as in their first derivatives. For this requirement to be satisfied, the following two conditions on the parameter values must be met:

$$T_A^3 = V_{THP}/\alpha \quad t_0 = (V_{DD} - V_{THP})/3\alpha T_A^2$$

The parameter β is to obtain flexibility in adjusting the rate of the output-pulldown process. If $\beta = 0$ [Fig.1.05.4(b)] the output-node pulldown process after $t = T_A$ is a simple time-constant process.

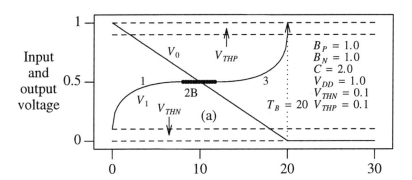

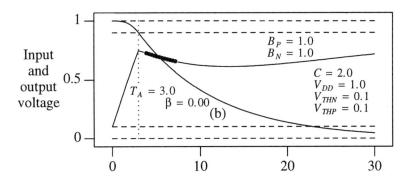

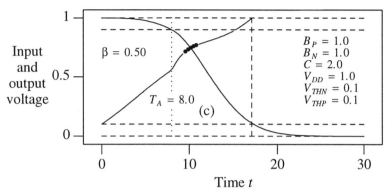

Figure 1.05.4 Trial runs of the new method of solution

The parameter values are given in the figure. In Fig.1.05.4(b), the V_1 waveform is not monotonically increasing with time. This can be explained as follows. The output node must pull down from V_{DD} to $V_{DD} - V_{THP}$ in a short time, $T_A = 3.0$. This requires that V_1 must increase to a high voltage level in a short time. The rate of decrease of V_0 after T_A (by the exponential dependence) is smaller than the rate at time $t = T_A$, and therefore V_1, which once reached a high level, must decrease to accommodate the smaller rate of decrease of the output-node voltage V_0 thereafter. The nonmonotonicity of V_1 originates from the V_0, which has the maximum negative slope at $t = T_A$, as is observed from the mathematical form of Eqs.(1.05.9a) and (1.05.9b). V_1 maximum occurs at the maximum of V_0 slope that is, at T_A. Strictly speaking, this interpretation is valid only if the pulldown NFET is still in the saturation region. For Fig.1.05.4(b) the assumption is, of course, valid.

Figure 1.05.4(c) shows an example having $\beta > 0$. The maximum of V_1 becomes higher than V_{DD} and moves out of the voltage domain. If $\beta > 0$, location of the maximum of V_0 slope (the negative steepest slope) moves to the right by increasing the rate of decrease of V_0 in the region $t > T_A$. $V_1(t)$ of Fig.1.05.4(c) reaches $V_{DD} - V_{THP}$ when the PFET turns off. In this figure, $V_1(t)$ is certainly continuous, but $dV_1(t)/dt$ is not, at $t = T_A$. This is because $V_1(t)$ includes $dV_0(t)/dt$ as an argument, and therefore $dV_1(t)/dt$ includes the second derivative $d^2V_0(t)/dt^2$. This can be seen as follows. To get $dV_1(t)/dt$, V_1 as a function of V_0 and of dV_0/dt is differenciated once by t. Then the resultant derivative of V_1 includes the second derivative of V_0. This second-order derivative is not continuous at $t = T_A$ by the $V_0(t)$ given by Eqs.(1.05.9a) and (1.05.9b). This example shows that to obtain reasonable (smooth and featureless) $V_1(t)$, sensitivity of $V_1(t)$ waveform on the assumed $V_0(t)$ waveform is kept in mind because the $V_1(t)$ algebraic expression includes a derivative of $V_0(t)$ as well as $V_0(t)$ itself. It is not a trivial matter to choose a $V_0(t)$ that gives a smooth, monotonous, featureless, and natural-looking $V_1(t)$. This sensitivity provides an insight into inverter-switching mechanisms and moreover, into the mathematical and physical bases of this method of solution. A problem formulated by a differential equation whose independent variable is time expresses causality. For a complete set of causes, the effect is uniquely determined. If the cause is rational, the effect looks natural. It is not possible, however, to assume that an effect that looks natural from a casual observation has always a rational cause. This is a crucial point in *backtracking* the causal relationship by this mathematical method. Hypersensitivity of the input waveform to the assumed output waveform is caused by inclusion of the output waveform derivative. One way to avoid an unnatural $V_1(t)$, $V_0(t)$ should be approximated by an analytic function of time t over the entire range of t, so that V_0 and dV_0/dt can be differentiated any number of times.

This is crucial mathematically, whether or not the multivariable function ΔI can be solved for V_1. If $\Delta I(V_{DD}, V_1, V_0)$ is an analytic function of the arguments over the entire range, whether or not V_1 can be solved is determined from a simple criteria, $\partial V_0 / \partial V_1 \neq 0$. Generally, the singular points where the solution becomes impossible are isolated points. If ΔI is nonanalytic as in Eqs.(1.05.2) and (1.05.3), then it may so happen that V_1 cannot be solved in a certain range. An illustrative example follows. If V_1 begins to increase from zero and if $V_1 < V_{THN}$, then the NFET is in the nonconducting region, and $V_0 = V_{DD}$ and $dV_0(t)/dt = 0$. In this range V_1 may take any voltage waveform and does not affect V_0 at all. In this range $V_1(t)$ cannot be determined from the given $V_0(t)$. This is another example of the peculiarity originating from backtracking causality. Metaphorically, it may be said that a *shadow* appears in the cause, if the problem is illuminated from the effect.

In Fig.1.05.4(c) the entire V_0 waveform was shown by the solid curve that is, the assumed $V_0(t)$. Once $V_1(t)$ reaches V_{DD} (to the vertical dashed line at $t = 17$), however, the assumed output waveform is not valid thereafter. In the later times the residual pulldown of the inverter is carried out autonomously or the input-output connection through the inverter is cut off, and the pulldown process becomes self-deterministic or the inverter goes into an autonomous pulldown. The process is described mathematically as follows. The circuit equation in this time zone is given by

$$C\frac{dV_0}{dt} = -B_N[(V_{DD} - V_{THN}) - (1/2)V_0]V_0$$

If we introduce $V_X = V_{DD} - V_{THN}$ and $U_0 = V_0 - V_X$ the equation is reduced to

$$C\frac{dU_0}{dt} = -(B_N/2)(V_X^2 - U_0^2)$$

which is solved by the separation of variables as

$$(1/2V_X)[\log(V_X+U_0)-\log(V_X-U_0)] = -(B_N/2C)t+const$$

If $V_0(t_C) = V_{0C}$ is the required initial condition we have a solution

$$V_0(t) = \frac{2(V_{DD}-V_{THN})V_{0C}\exp[-(t-t_C)/t_0]}{2(V_{DD}-V_{THN})-V_{0C}[1-\exp[-(t-t_C)/t_0]]} \quad (1.05.10)$$

In Fig.1.05.5, the input-controlled domain and the autonomous time region of the inverter pulldown process are shown. In Fig.1.05.5 in the autonomous region, the assumed $V_0(t)$ becomes invalid. This is yet another feature of backtracking causality: the assumption made on the output waveform is violated as the system evolves and arrives at the autonomous region. In the autonomous region V_0 is shown by the dotted curve, which is a plot of Eq.(1.05.10).

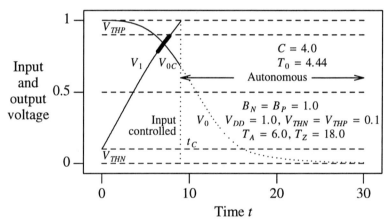

Figure 1.05.5 Separation of $V_0(t)$ to input-controlled and autonomous domains

As we discussed before, $V_0(t)$ used to compute $V_1(t)$ should be an analytic function of t. Then what are its requirements?

(1) Since we consider a pulldown process,

$$V_0(0) = V_{DD} \quad V_0(\infty) = 0$$

must be satisfied.

(2) At the beginning of the pulldown process the circuit equation is

$$C\frac{dV_0}{dt} = B_P[(V_{DD}-V_{THP}-V_1)-(1/2)(V_{DD}-V_0)](V_{DD}-V_0)$$

$$-(1/2)B_N(V_1-V_{THN})^2$$

At the moment when increasing V_1 reaches V_{THN} and increases beyond that, $V_1(t)$ can be expanded into a Taylor series and only up to the first t-dependent term is retained. We obtain

$$V_1(t) \approx V_{THN} + \gamma t$$

where γ is a constant, and where zero of time t is chosen at the moment when V_1 equals V_{THN}. If $V_0(t)$ is approximated by $V_0(t) = V_{DD} - \alpha t^n$ in the vicinity of $t = 0$, the circuit equation in the limit as $t \to 0$ gives

$$-nC\alpha t^{n-1} = -(1/2) B_N \gamma^2 t^2 + o(t^3)$$

To satisfy this equation we have $n = 3$ and

$$V_0(t) \to V_{DD} - \alpha t^3 \quad (t \to 0) \quad \text{where} \quad \gamma^2 = (6C/B_N)\alpha$$

(3) In the other limit, if V_1 increases beyond $V_{DD} - V_{THP}$ and the PFET is turned off, the output node is pulled down by the NFET only. In this regime, the circuit equation is written as

$$C\frac{dV_0}{dt} = \Delta I = -B_N[(V_1 - V_{THN}) - (1/2) V_0] V_0$$

and in the limit as $V_0 \to 0$, the equation is further simplified by neglecting the V_0^2 term, to a linear differential equation that has a simple exponential solution:

$$C\frac{dV_0}{dt} = -B_N(V_1 - V_{THN}) V_0$$

From the equation, V_0 is a time-constant process described by $V_0 \approx const. \exp(-t/t_0)$ in the limit as $t \to \infty$ where the time constant is given by

$$t_0 = C/[B_N(V_{DD} - V_{THN})]$$

It is interesting to note that if t_0 were chosen arbitrarily, the input voltage limit $V_1(\infty)$ derived from the given output node voltage $V_0(t)$ is not V_{DD}. It is V_{DD} only if $t_0 = [C/[B_N(V_{DD} - V_{THN})]]$. This requirement is, however, less strictly enforced than (1) or (2), because of the assumption $V_1 \approx V_{DD}$ in the last phase of the pulldown process. In this domain the inverter goes into the autonomous zone. Yet this condition is desirable, since this guarantees smooth connection to $V_0(t)$ to the autonomous regime. Requirement (3) represents the inverter in the last phase of the pulldown transition, when the inverter goes into the deep autonomous regime. The dependence is called an *exponential tail*. As we see, the $V_0(t)$ must satisfy many conditions (1) to (3) just to ensure that $V_1(t)$ derived from it does not look absurd. Since the inverter determines $V_0(t)$ from $V_1(t)$, solution $V_0(t)$ exists for any given $V_1(t)$ uniquely. What we are doing here is not what nature does. If $V_0(t)$ is given instead, $V_1(t)$ may not exist, or the mathematical solution may not be physically acceptable. Solution $V_0(t)$ is quite insensitive to $V_1(t)$, but solution $V_1(t)$ is quite sensitive to $V_0(t)$, and this is an interesting feature of this method. Yet discoveries in physics have been accomplished by backtracking causality. There certainly was good luck in finding universal gravity, since the orbit of the planet was an ellipse. Backtracking the effect to the cause, the force proportional to the inverse square of distance was discovered. The discovery of energy quanta was less fortunate: the best empirical formula of black-body radiation at that time did not lead to the concept of quantized energy.

The overall $V_0(t)$ waveform in a pulldown process can be expressed by a simple exponential term, $V_0(t) = V_{DD} \exp(-t/t_0)$. The formula satisfies conditions (1) and (3). The simple expression does not satisfy condition (2), or the required t-dependence in the limit as $t \to 0$. To correct this, it is convenient to stretch or shrink the time scale by using a transformation on t. Then we set

$$V_0(t) = V_{DD} \exp[-f(t)/t_0]$$

If we choose

$$f(t) = t^3/(t^2 + at + b)$$

all the limits are satisfied. In the limit as $t \to 0$, $f(t) \to t^3/(bt_0)$, and then we are able to choose b by $\alpha = 1/(bt_0)$. Parameter b determines the overall delay of the output waveform to initiate the pulldown process relative to $t = 0$ (the time when the increasing input voltage reached V_{THN} to initiate the pulldown transition). At about $t = \sqrt{b}$, the time dependence of $f(t)$ changes from $f(t) \approx t^3$ ($t \to 0$) to $f(t) \approx t$ ($t \to \infty$). The function is analytic over the entire domain of t, and therefore the problem of connecting functions from different regimes can be avoided. Figure 1.05.6(a)-(e) shows the input and the output waveforms for a typical set of the parameter values computed using the trial function, $V_0(t)$, covering in order the dynamic to the static switching regimes. We note that once $V_1(t)$ reaches V_{DD}, $V_0(t)$ goes into the autonomous region, and the assumed waveform becomes invalid. On the right side of the vertical dotted lines of Figs.1.05.6(a)-(d) the solution Eq.(1.05.10) is connected. Condition (3) of $V_0(t)$ is not mandatory.

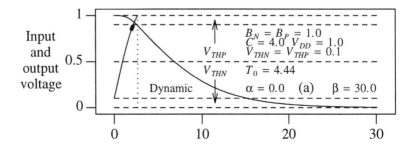

Classical Gate Field

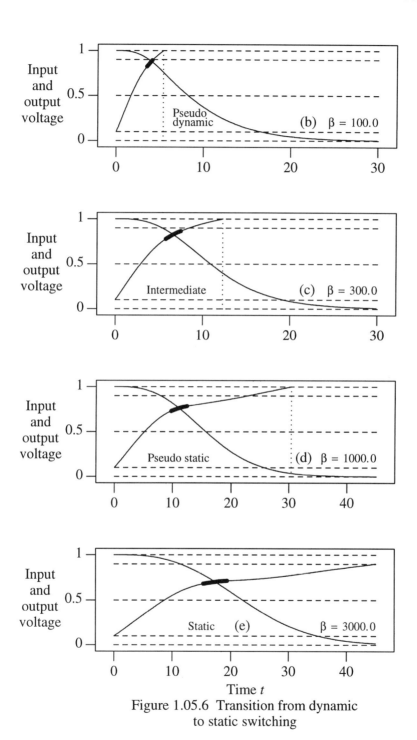

Figure 1.05.6 Transition from dynamic to static switching

This condition, however, guarantees that $V_1(t)$ is rational under a wide range of conditions, especially in the deep static switching regime shown in Fig.1.05.6(d) and (e).

From these examples a question arises: why are normally observed digital-node waveforms quite featureless? There is a mechanism for eradicating the features of the node waveforms. The dynamic switching mode is the dominant mode in which a digital circuit operates. In a dynamic switching mode, input to output connection is cut off once the inverter goes into the autonomous region. This is reflected in the mathematics of the causality backtracking method: the assumed $V_0(t)$ becomes invalid in the autonomous regime. In the autonomous regime the waveform that reflects the simple internal structure of the inverter given by Eq.(1.05.10), which does not depend on the complex input waveform, is displayed. Certain circuits, like a ringoscillator, work only in the dynamic switching mode (see Shoji, 1992).

Figures 1.05.6(a) to (e) show the transition from the dynamic to the static switching mode. In this examples, $a = 0$ and $C = 4$ for all. The parameter b was increased from (a) to (e). For rapid input switching from 0 to V_{DD} at small b, the inverter goes into the autonomous regime almost immediately, and after that the input waveform does not matter. In the dynamic switching mode, the choice of the output waveform does not affect the input waveform significantly. As parameter b is increased, transition to the static switching mode takes place as follows. For larger b, to maintain the same inverter parameters consistent with ever-increasing output delay and fall times, $V_1(t)$ must increase ever more slowly, thereby increasing the transition time of the input waveform. Then the inverter stays a long time in the region where the output is susceptible to the input. Then the choice of the output-waveform reflects back to the input-waveform sensitivity.

The mathematical method of analysis of a CMOS inverter can be used to study complex inverter problems that involve both input and output waveforms, such as the overlap current and the dependence of the inverter delay time on the input signal rise and fall times. Figure 1.05.7(a)-(c) shows the overlap current waveforms determined by the causality backtracking method. The overlap current of the inverter of Fig.1.05.1(a) during the output pulldown transition equals the PFET current I_P. The current is a waste of the pulldown process, since the pulldown can be carried out by the conducting NFET alone. If the PFET turns off immediately after the start of the input pullup, the wasted current and the energy are zero. Since the input rise time is not zero, there is a period of time when both PFET and NFET are turned on. During that time current flows directly from V_{DD} to ground. Figures 1.05.7(a)-(c) show the input and the output voltage waveforms and ten times the overlap current waveform, for successively longer input rise times. The degree of the overlap current effects can be measured by the ratio of the overlap charge defined by

Classical Gate Field 35

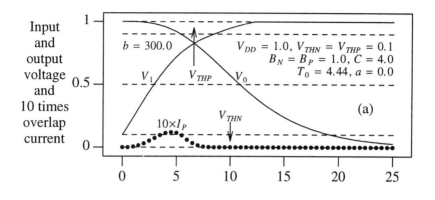

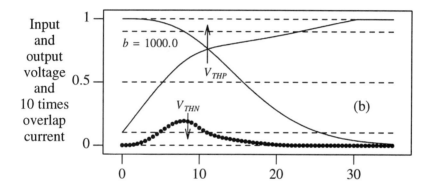

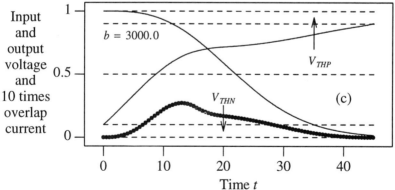

Figure 1.05.7 Overlap current of CMOS inverter

$$Q_O = \int_0^\infty I_P(t)\,dt$$

and by the charge stored in the load capacitor, CV_{DD}. The input rise time is defined by the time t_1 required for $V_1(t)$ to increase form

$V_1(0) = V_{THN}$ to $V_1(t_1) = V_{DD}/2$. Figure 1.05.8 shows $Q_O/(CV_{DD})$ versus the input rise time t_1, for various values of the load capacitance C. The normalized overlap charge increases with increasing input rise time and with decreasing load capacitance, and both are reasonable.

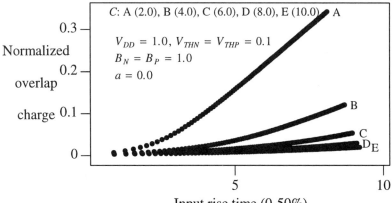

Figure 1.05.8 Inverter charge loss versus input rise time

Another application of the causality backtracking method is to determine the dependence of the inverter delay time versus the input rise time. The inverter delay time is defined as the interval from the time when the increasing V_1 reached V_{LOGIC} to the time when the decreasing V_0 reaches V_{LOGIC}. V_{LOGIC} is the common decision threshold of the logic level. Choice of V_{LOGIC} is uncertain, however, and the uncertainty becomes the central issue in Chapter 2. Here, $V_{LOGIC} = V_{DD}/2$ is a choice, since the inverter is pullup and pulldown symmetrical (the classical delay definition). Figure 1.05.9(a) shows this case. As the input rise time increases, the delay time increases as well. If the input rise time is longer, the effective input voltage during the period when $V_0 > V_{LOGIC}$ (= $V_{DD}/2$) is lower. Accordingly, the conductivity of the NFET is lower, and the pulldown transition takes longer.

If the inverter is not pullup and pulldown symmetrical, how to set V_{LOGIC} becomes ambiguous. If we set V_{LOGIC} still at $V_{DD}/2$, we get results shown in Fig.1.05.9(b) (for stronger pulldown, $B_N > B_P$) and in Fig.1.05.9(c) (for stronger pullup, $B_N < B_P$). As shown in Figs.1.05.9(a) to (c), dependence of the inversion delay on the input rise time is almost linear.

The mathematical method of causality backtracking can be generalized to more complex circuits, such as a two-stage cascaded inverter analysis. The node voltages of the successive nodes, V_1, V_A, and V_B satisfy the circuit equation

$$C_A \frac{dV_A}{dt} = \Delta I_A(V_{DD}, V_1, V_A) \quad C_B \frac{dV_B}{dt} = \Delta I_B(V_{DD}, V_A, V_B)$$

By giving a trial function $V_B(t)$, $V_A(t)$ can be determined from the second equation, and from $V_A(t)$, $V_1(t)$ is determined from the first equation. There are two issues that require attention, however. In the region where V_1 affects V_B directly, V_1 can be solved from the two equations as

$$V_1 = \Psi[V_A[V_B,(dV_B/dt)],(d/dt)V_A[V_B,(dV_B/dt)]]$$
$$= \Phi[V_B,(dV_B/dt),(d^2V_B/dt^2)]$$

and the expression contains the second derivative of V_B.

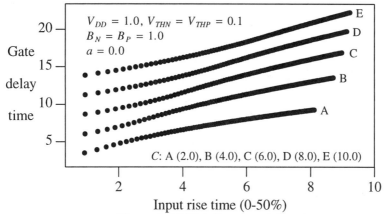

Figure 1.05.9(a) Inverter delay time versus input rise time

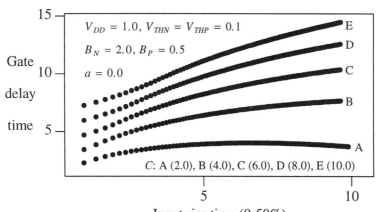

Figure 1.05.9(b) Inverter delay time versus input rise time

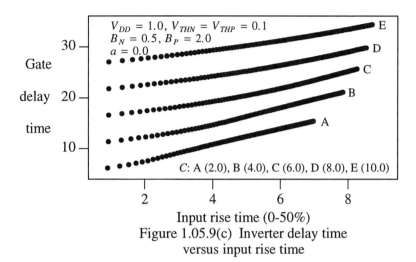

Figure 1.05.9(c) Inverter delay time versus input rise time

This means that the sensitivity to the selection of V_B is even more pronounced than the single-stage inverter: If V_1 is to be smooth, the third derivative $d^3 V_B(t)/dt^3$ must be continuous.

As we saw before, a $V_A - V_B$ relationship separates into the two regions, where V_B is autonomous and where it is not. If the V_A solution is used to determine V_1, the $V_1 - V_A$ relationship separates into the similar two regions. Then, if they are in the region where V_A becomes autonomous, V_A and V_B become inconsistent, and the original selection of V_B must be modified. Again, this is the issue of sensitivity of the input waveform to the choice of the output waveform, the ideosyncracy originating from backtracking causality.

From the study of the inverter-switching problem, we learned the following. (1) Physically meaningful input waveforms do not exist for every conceivable output waveform: (2) The input waveform determined from the output waveform contains the higher derivatives of the output waveform and is sensitive to the output waveform selection: (3) In certain cases the system becomes insensitive to the input change, and an area that may be called a *shadow* is created, where the input waveform cannot be determined from the output waveform: and (4) In some region the assumed output waveform may become inconsistent with the output waveform that is reconstructed from the input waveform, which is determined by backtracking the originally assumed output waveform. Then the output waveform must be modified, in the contradicted region, which is the autonomous region. The four featuress of this mathematical technique are considered quite common feature of the causality backtracking method based on mathematical grounds, and they are expected to be observed in any case when the method is used. In spite of the peculiarities, problems can be solved in a closed form, and any mathematical difficulty in the procedure seems to reflect the system's own peculiarity.

Classical Gate Field

Backtracking causality seems to clarify the information flow in the physical system in great detail and highlights the structure of the information flow. In some problems, observing nature in reverse may be the better way to understand it, rather than following the natural causal evolution. If this conjecture turns out to be correct, this method would be useful in studying many physical systems.

1.06 The Velocity of the Propagation of Excitation

Energy transfer in a gate-field allows the observer to interpret the movement of an identifiable object, an excitation. An excitation has its own identification, and therefore it can be associated with moving information. It has many attributes of a particle, both the classical mass point and the elementary particle of quantum field theory. The most important parameter is its velocity in the gate-field.

Let us consider an inverter field. The left end of an infinitely long and uniform inverter field was at the LOW logic level for a long time to establish a steady state in the field, and then input node is pulled up at $t = 0$. After some time the input node reaches the HIGH logic level. The node voltage waveform that propagates down the gate-field depends on the input waveform in the immediate vicinity of the input end. After some stages of propagation, however, the node waveform approaches a steady-state waveform that satisfies

$$V_{n+2}(t) = V_n(t - 2\tau) \quad \text{or} \quad (1.06.1)$$
$$V_{n+1}(t) = V_{DD} - V_n(t - \tau)$$

The first condition holds generally in any uniform gate-field. The second, a stronger condition, holds in the pullup and pulldown symmetrical gate-field. An inverter field becomes simple if it is pullup-pulldown symmetrical. If the input and output voltage waveforms of an inverter are $V_1(t)$ and $V_0(t)$, respectively, and if the inverter is pullup-pulldown symmetrical, the following time-axis translational symmetry emerges. Upon the input waveform transformation

$$V_1(t) \rightarrow V_{DD} - V_1(t)$$

the output waveform of a symmetrical inverter is transformed to

$$V_0(t) \rightarrow V_{DD} - V_0(t)$$

We assume the pullup-pulldown symmetry for simplification, and we try to get some idea about how to define the velocity of excitation. When the steady propagation state of the wavefront is reached, the velocity of propagation of the excitation can be defined by time shift and by matching of the node waveforms. This precise definition of excitation velocity is consistent with the definition of a particle's velocity in classical mechanics. Since the excitation takes time τ to propagate one stage, its velocity v is defined by

$$v = 1/\tau \text{ (stage/second)}$$

Delay time τ and voltage profile $V_n(t)$ are determined by seeking for consistency from the set of the circuit equations. Delay time τ is the eigenvalue. The eigenvalue problem is solved later in Sections 2.04 and 2.05. Here it suffices to say that excitation velocity can be unambiguously defined.

Let us study how the velocity depends on the gate-field parameters. We gain insight through a computer experiment. Let the FETs of the inverter field be modeled by the gradual-channel, low-field FET model. The NFET current I_N is given by

$$I_N = B[V_1 - V_{TH} - (1/2)V_0]V_0$$
$$(V_1 > V_{TH}) \quad \text{and} \quad (V_0 < V_1 - V_{TH})$$
$$I_N = (B/2)(V_1 - V_{TH})^2 \quad (V_1 > V_{TH}) \quad \text{and} \quad (V_0 > V_1 - V_{TH})$$

and $I_N = 0$ if $V_1 < V_{TH}$. A similar relationship for the PFET current I_P is given by replacing

$$V_0 \to V_{DD} - V_0 \quad \text{and} \quad V_1 \to V_{DD} - V_1$$

Pulldown switching of an inverter begins at the moment when the NFET is turned on, and in the early phase of the pulldown the NFET is in the saturation region. If the input switching is fast (dynamic switching), the input-voltage may be approximated by V_{DD}, neglecting the input-voltage transition region. Then the NFET current I_N is the saturation current, $I_{SAT} = (B/2)(V_{DD} - V_{TH})^2$. The time required to discharge the output node charge $Q = CV_{DD}$ is

$$T_D = CV_{DD}/I_{SAT}, \quad \text{or} \qquad (1.06.2)$$
$$\text{velocity} \approx B(V_{DD} - V_{TH})^2/CV_{DD}$$

This estimate must be modified by including a factor of the order of unity to make the best fit to the real excitation velocity. Figure 1.06.1 shows the results of correlating the parameter to the velocity. By seeking for the best match we obtain

$$\text{velocity} = 1.11075 B(V_{DD} - V_{TH})^2/CV_{DD} = \phi \qquad (1.06.3)$$

This formula shows that the velocity of propagation of an excitation is determined by the parameters of the uniform gate-field alone. A propagating excitation in a uniform field appears as if it were a mass point moving in a space that exercises no force on it. The law of inertia in mechanics holds. The velocity of the excitation changes if the field parameters vary. In Fig.1.06.2, velocity of propagation of an excitation in a nonuniform gate-field that has locally decreased propagation velocity is shown. The FET conductance parameter B is reduced from 1.0 to 0.5 in a notch seven stages long.

Classical Gate Field

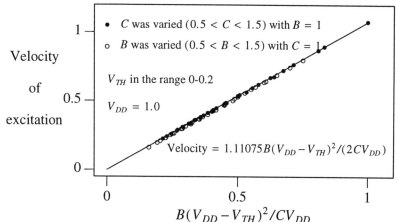

Figure 1.06.1 Dependence of steady-state velocity versus the delay parameter

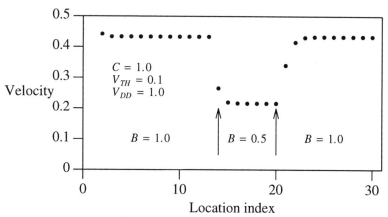

Figure 1.06.2 Velocity versus location

The location of the gate-field where the conductance parameter B is changed is shown by the arrows. Strictly speaking, the velocity of an excitation whose voltage profile changes cannot be defined as cleanly as before, and this is a problem that is investigated later in Chapter 2. In Fig.1.06.2 the velocity is defined by the crossing of the half voltage point ($V_{DD}/2$) between the HIGH and the LOW logic voltage levels, since the gate-field is still everywhere pullup-pulldown symmetrical. In this case the definition is consistent with the rigorous time shift-match method described before. Use of the $V_{DD}/2$ logic threshold in the CMOS gate-delay definition is popular (Shoji, 1987), but the definition requires some interpretation. This discussion is postponed to Chapter 2. The propagation velocity changes and quickly settles at the new level determined by the notch-region gate-field parameter. This can be explained by assuming that a force is exercised to excitation at the location of the arrows, which accelerate or decelerate the excitation. The force is proportional to the difference of the parameter that determines velocity (defined before), ϕ, which is analogous to the potential energy of a particle in classical mechanics. Figure 1.06.2 shows that the velocity of excitation changes at the field parameter discontinuity but not instantly. Velocity of excitation takes a few stages of propagation before adjusting itself to the new-steady state velocity. Figure 1.06.3 shows how many propagation stages are required to settle at the final, steady-state velocity. It takes only one to two stages, at most, to settle at the new velocity level. Let us remember that the velocity of excitation in a gate-field is the eigenvalue of the eigenvalue problem, and that the eigenvalue is stable against the selection of a trial eigenfunction, if the eigenvalue problem is solved by the variational method (Landau and Lifshits, 1958). The circuit equation that defines the eigenvalue problem is

$$C\frac{dV_0(t)}{dt} = \qquad (1.06.4)$$
$$B[V_0(t-\tau) - V_{THP} - (1/2)[V_{DD} - V_0(t)][V_{DD} - V_0(t)]$$
$$-(B/2)[V_{DD} - V_0(t-\tau) - V_{THN}]^2$$

The eigenvalue τ is sensitive to B and C but not so much on the selection of the trial function $V_0(t)$.

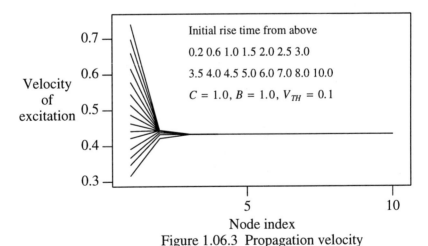

Figure 1.06.3 Propagation velocity versus node index

To determine the propagation velocity of excitation, it is necessary to formulate a mathematical representation of the switching process of a gate in a simple and manageable closed form that has clear physical meanings. This has not been possible because of the mathematical difficulties originating from the differential equation describing the circuit, as was discussed in Section 1.05. Instead, we make more observations of the mathematics and of the numerical results, and we try to draw useful conclusions. The first issue is how to define excitation velocity in classical analysis.

The propagation velocity of excitation in a uniform, symmetrical gate-field has been explained by simple waveform matching. The definition of the eigenvalue problem by a functional equation

$$V_n(t) = V_{DD} - V_{n-1}(t-\tau)$$

where τ is the delay time of an inverter, can be used to determine the excitation velocity. We proceed to study the velocities of excitation in uniform but asymmetrical gate-field, and in a locally nonuniform gate-field. If the field is uniform but is asymmetrical for pullup and pulldown, the steady-state propagation velocity of the excitation is determined over two inverter stages from the relationship

$$V_n(t) = V_{n-2}(t-\tau')$$

where τ' is the cumulative delay time of the two stages. The gate-field may be considered as cascaded noninverting buffer, each of which is made by cascading two stages of identical but asymmetrical inverters. The propagation velocity can be defined clearly only if the two inverter stages are considered together. Suppose that the PFET of the inverter is smaller than the NFET, or $B_P < B_N$. Then if the input of the cascaded buffer is pulled up, the output of the first stage pulls down soon after the input transition, but the output of the second stage (the output of the noninverting buffer) takes some time to pull up. The propagation velocity goes up and down within a unit stage of a noninverting buffer, as shown by the × sign of Fig.1.06.4, but if many of them are cascaded, the overall velocity v (per stage) is the harmonic average of the high (v_1) and the low (v_2) velocities as

$$(2/v) = (1/v_1) + (1/v_2) \tag{1.06.5}$$

The velocity oscillates as shown by the × signs of Fig.1.06.4.

If the two stages of the inverters are scaled such that the first stage has small B_P and large B_N, and the second stage has large B_P and small B_N, the noninverting buffer responds rapidly to the input pullup transition but slowly to the input pulldown transistion. The excitation velocity for the input upgoing transition is faster than that of the input pulldown transition. In Fig.1.06.4 closed circles and open circles show the velocities of the input pullup transition and pulldown transition, respectively. In an asymmetric but uniform field the two velocities were the same. In a locally nonuniform field the two velocities become different. There is a close parallel to the propagation of a longitudinal wave and a transverse wave in solids.

Summarizing the observations of this section, the two features of the velocity of excitation are as follows: (1) In a uniform but asymmetric gate-field, the overall propagation velocities of the two polarities of transition are equal, but if it is observed locally, it oscillates: (2) In a locally nonuniform but globally uniform gate-field the velocities of propagation of the input upgoing transition and of the input downgoing transition become different. The three cases of excitation-velocity measurement by the waveform matching is rigorous, yet they are applicable only to special cases. The method requires the capability of precision analog voltage measurement, which is not allowed in the quantum-mechanical modeling discussed in Chapter 2. Less rigorous, but more flexible delay measurement method is discussed below.

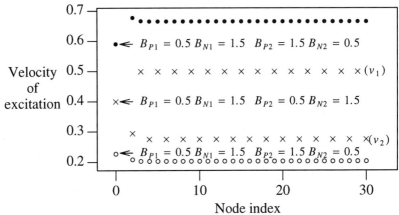

Figure 1.06.4 Propagation velocity of asymmetrical inverting buffer field

Delay of the inverter has been defined by setting the logic threshold voltage at $V_{DD}/2$. If the rigorous waveform time-shift-matching is not possible, setting a logic threshold is the only way. The level setting depends on the viewpoint where is the *center* or the representative coordinate of the excitation? since it is spread out over space and time. A similar problem exists in determining the velocity of a solid body. Where in a solid body should the location of the body be chosen? In classical mechanics, the center of gravity can be defined, and the body's motion can be separated into the motion of the center of gravity and the rotation around the center. From this viewpoint the oscillating excitation velocity of the uniform but asymmetrical gate-field may be considered because the coordinate of the excitation is set not at the right place, and therefore extra complications arose. A logic threshold can be set to make the velocity of the two cascaded stages equal. Such a level may be regarded as defining the *center of gravity* of the excitation, yet still this is an arbitrary choice.

The issue of the center of an excitation has a fundamental

implication in modeling digital-circuit operation. For an excitation spreading out over space to be considered to be a solid body like object, the physical parameters at every point within the body must be measurable. A signal in a digital circuit observed by a cathode-ray tube (CRT) oscilloscope appears like a classical object that is solid, but actually it is able to deform. This simple solid body like model, however, cannot be pushed all the way in the digital circuit theory, and this is the source of problems and, often, confusion. The problem is most clearly seen in the impossibility of setting a rational logic-threshold voltage. This difficulty necessitates the introduction of a quantum-mechanical interpretation into the theory. In the quantum-mechanical interpretation the spatial extension of excitation is interpreted by probability distribution. The center is then the expectation value of the spatial coordinate. The physical meanings of the center of gravity and the expectation value are fundamentally different. In a digital circuit each gate is a measurement setup of the Boolean-level. A NAND gate and a NOR gate determine location or the arrival time of an excitation differently, since their logic threshold voltages are not the same. All these determinations must be valid for the logic circuit to work correctly, and since we do not know the gate types at a particular location, a logic-threshold voltage is a very ill-defined parameter.

1.07 An Equation of the Motion of Excitation

An isolated excitation propagating in a gate-field maintains its identity, although some ambiguity about its location exists. This is the essential attribute of a mass point in classical mechanics. Then its motion must be governed by an equation of motion. I derive the dynamic equation of motion in this section. Let the nodes of an one-dimensional gate-field be indexed by n, a discrete variable used to identify location on it. If the excitation arrives at location $(n-1)$ and n at t_{n-1} and t_n, respectively, the velocity of the excitation v at location n is defined by

$$v(n) = [n-(n-1)]/(t_n - t_{n-1}) = 1/(t_n - t_{n-1}) \quad (1.07.1)$$

In this definition, it is assumed that the times t_n and t_{n-1} can be determined unambiguously. This is not generally true. This issue is related to whether we consider the gate-field a classical object or a quantum-mechanical object. In this section we consider a *classical* gate-field, in which the problems of quantum-mechanical ambiguity, which are discussed in Section 2.09, are ignored. This means that the logic-threshold voltage can be set *uncontrovertibly* at a certain level. Subject to this condition, the equation of motion describes the dynamics of a classical digital excitation.

The velocity is a function of location index n. Since the excitation moves with time, the location of excitation depends on time. We indicate the dependence by $n(t)$, where $n(t)$ and t are related by the equation

$$t = \int_0^{n(t)} \frac{dn}{v(n)} \quad \text{or} \quad t = \sum_{n=0}^{n(t)} \frac{1}{v(n)} \quad (1.07.2)$$

where we assume $n(0) = 0$. This is the formula that gives delay of a cascaded gate chain as a sum of the delays of the individual gates.

If the gate-field is uniform, the velocity of excitation is determined by the parameter ϕ, defined in Section 1.06, which is a function of the parameters specifying the gate-field at the node location n. The velocity is then written as $v_0[\phi(n)]$ as a function of node index n. The velocity of excitation in an approximately uniform gate-field is given by v_0, subject to the condition

$$|\phi(n) - \phi(n-1)| \ll |\phi(n)|$$

According to the numerical analysis given in the last section, if v_0 changes suddenly from one location to the next, the velocity of excitation is unable to settle at the new velocity instantly. The transient process is described approximately by an exponential decay to the new velocity with respect to location index n. We define the error of approximating v by v_0 by small deviation Δv, defined by

$$\Delta v = v(n) - v_0[\phi(n)] \quad (1.07.3)$$

The exponential decay observed from the numerical-analysis result is then described by

$$\frac{d\Delta v}{dn} = -\alpha \Delta v \quad (1.07.4)$$

where α is a factor that determines the speed of reestablishing the new steady-state velocity. Let us consider the physical meaning of the factor α.

Suppose that an excitation propagates along a uniform and symmetrical gate-field. If we observe at a gate location, the input and output waveforms are consistent, in that the output waveform becomes the input waveform of the next stage, thereby recreating the output waveform at its output location once again. The parameters characterizing the waveform, such as the rise-fall times of the excitation, are thus determined self-consistently from the uniform parameter values of the gate-field. Suppose that the steady-state velocity changes suddenly at a location. If the left side has higher velocity than the right side, the input waveform to the gate at the location has shorter rise time than the output fall time. The output fall time is, however, an attribute of the gate at the location, and that does not decrease significantly, even if the input rise time is reduced to zero. This is because the significant part of the delay time is determined by the autonomous regime of switching, which was discussed in

Classical Gate Field

Section 1.05. By going through one stage of gate, the rise-fall time is corrected quite significantly. If the left-side velocity is lower than the right-side velocity, the gate input voltage increases slowly, and therefore the gate goes through a quasi-static switching. Since a gate has high small-signal gain at the switching threshold, the slow change of the input waveform is amplified, and the output fall time is quite significantly reduced by going through only one stage. Here it is crucial to note that if the input rise-fall time is longer, the gain is higher because a digital gate operates at the gain-bandwidth limit (Shoji, 1996). From this consideration, α of Eq.(1.07.4) is a nondimensional number of the order of unity. By observing Fig.1.06.3 we may set $\alpha = 1$ as the first approximation.

By combining the equations we obtain

$$\frac{dv(n)}{dn} = \frac{dv_0(n)}{dn} - \alpha[v(n) - v_0(n)] \qquad (1.07.5)$$

If we observe the dynamics of excitation from itself, $v(n)$ is replaced by a function of t and is written as $v(t)$, and n is given by Eq.(1.07.1). $v_0(n)$ is the function of location n. Then the dynamic equation is written by multiplying $v(t) = dn(t)/dt$ as

$$\frac{dv(t)}{dt} = \frac{dv_0(n)}{dn} v(t) - \alpha v(t)[v(t) - v_0(n)] \qquad (1.07.6)$$

The equation consists of the two terms: the first term on the right side is a *force* exercised on the excitation from the gate-field, and the second term is the relaxation effect. The first term is the gradient of a potential, $v_0[\phi(n)]$.

The relaxation effect is the key feature of the dynamic equation. The effect is caused by the dependence of the inverter delay on the input signal transition rate. The important issue is why the excitation velocity tends to the steady-state velocity of the uniform field so quickly. To gain more insight into this point and to gain more confidence about the equation, let us analyze a model that has a closed-form solution. Let us consider a nonuniform, pullup-pulldown symmetrical inverter chain. Let the FETs of the chain be modeled by the collapsable current generators with zero-conduction threshold voltages. The transconductance of the FET is G_m. Since this problem has been solved before (Shoji, 1992), only the outline of the analysis is presented. The input and output node-voltage waveforms are schematically shown in Fig.1.07.1. The nodes 1 and 0 are loaded by capacitance C_α and C_β, respectively. The node waveforms in the regions bounded by the closed circles A, B, C and D are given, respectively, by

A: $V_1 = V_{DD} - \alpha t^2 \quad (0 \le t \le t_1)$

B: $V_1 = (V_{DD} - \alpha t_1^2) - 2\alpha t_1(t - t_1) \quad (t_1 \le t \le t_3)$

Y: $V_1 = 0 \quad (t_3 \le t)$

C: $V_0 = \beta(t-t_2)^2 \quad (t_2 \leq t \leq t_3)$

D: $V_0 = \beta(t_3-t_2)^2 + 2\beta(t_3-t_2)(t-t_3) \quad (t_3 \leq t \leq t_5)$

Z: $V_0 = V_{DD} \quad (t_5 \leq t)$

Curve V_1 connects smoothly at $t = t_1$, and curve V_0 connects similarly at $t = t_3$. At $t = t_1$, the input voltage to the first stage inverter reaches V_{DD}. To maintain the integrity of the analysis using the segmented waveforms, we assume that level $V_{DD}/2$, the logic threshold we use, is crossed only by the linear section of the curves and not by the parabolic section. Then we have

$$V_{DD} - \alpha t_1^2 > (1/2)V_{DD} \quad \text{or} \quad (1.07.7a)$$

$$2\alpha t_1^2 < V_{DD} \quad \text{and}$$

$$\beta(t_3 - t_2)^2 < (1/2)V_{DD} \quad \text{or} \quad (1.07.7b)$$

$$(G_m/C_\beta) < [(4\alpha t_1)/V_{DD}]$$

t_2, t_3, and t_4 are calculated as

$$t_2 - t_1 = \frac{(V_{DD}/2) - \alpha t_1^2}{2\alpha t_1}$$

$$t_3 - t_1 = \frac{V_{DD} - \alpha t_1^2}{2\alpha t_1} \quad \text{or} \quad t_3 - t_2 = \frac{V_{DD}}{4\alpha t_1}$$

$$t_4 - t_3 = \frac{(V_{DD}/2) - \beta(t_3 - t_2)^2}{2\beta(t_3 - t_2)}$$

The delay time T_D defined at the logic-threshold voltage $V_{DD}/2$ is given by

$$T_D = t_4 - t_2 \quad (1.07.8)$$

$$= \frac{(V_{DD}/2)}{2\alpha t_1} + \frac{(V_{DD}/2) - \beta(t_3 - t_2)^2}{2\beta(t_3 - t_2)}$$

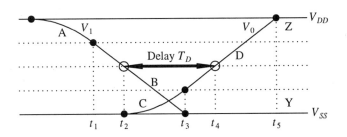

Figure 1.07.1 Node waveforms

In the linear region of the node waveform C and D of Fig.1.07.5 the PFET of the preceding stage and the NFET are turned off. Then we have

$$2\alpha t_1 = (G_m V_{DD})/C_\alpha \tag{1.07.9a}$$

$$2\beta(t_3 - t_2) = (G_m V_{DD})/C_\beta \tag{1.07.9b}$$

If the two relations are substituted, we have

$$T_D = (C_\alpha/4G_m) + (C_\beta/2G_m) \tag{1.07.10}$$

In the delay formula the first term is the contribution from the preceding stage, and the second term is the contribution from the present stage. The second term is the delay of the stage if the input signal makes an instantaneous transition.

The structure of the delay formula supports the assumption used to derive the dynamic equation of the excitation. Subject to the condition of waveform integrity, Eqs.(1.07.7a) and (b), the delay of an inverter stage depends on the parameters of the stage and to a lesser extent on the parameter of the preceding stage, but no other stage is involved. The reason that the range of influence is short is that an inverter does not transmit information once if it settles to a Boolean-level. In the autonomous region of switching discussed in Section 1.05, the inverter isolates the input from the output. Nonlinearity of inverter is the cause of the isolation. This is the reason that the excitation velocity settles quickly to the new steady-state value at the discontinuity of the gate-field parameters, as shown in Fig.1.06.3. To understand this problem, applying the concept of information flow in the circuit of Section 1.05 is quite useful.

1.08 Node Waveform of Logic Circuits

In the textbooks of logic-circuit design and analysis, the voltage waveforms of the logic nodes are shown, conventionally, by crisp, zero rise-fall time pulses, such as are shown in Fig.1.08.1(a) (Mano, 1991; Nagle, Carrol and Irwin, 1975). If the delay time of the logic circuit is included in the analysis, the crisp pulses are shifted to the right by the delay time, as shown in Fig.1.08.1(b). Various complicated problems in the circuit level are avoided by assuming the crisp waveforms, and the definition of the gate's delay time becomes clear and definite. A logic diagram is designed assuming the node waveforms of Figs.1.08.1(a) and (b). The real node waveform, however, never looks like the idealized crisp pulse: the pulse's rise-fall times occupy a significant fraction of the pulse width, as shown in Fig.1.08.1(c). The assumption that a logic gate generates a delayed crisp pulse from a crisp input pulse contradicts circuit theory. The delay time and the rise-fall time are strongly correlated (Shoji, 1987). If the delay time is defined at the logic-threshold voltage halfway between the HIGH and the LOW voltages, the delay time T_D is approximated by

$$T_D = (1/2)T_{R/F} + T_{TR}$$

where $T_{R/F}$ is the rise-fall time of the gate's output node, which is determined primarily by the output-impedance of the gate and the load capacitance of the output node. T_{TR} is the transit time of the signal from the input to the output of the gate. The physical meaning of T_{TR} is the time required to establish the final output-impedance level of the gate. If two FETs are connected in a series, T_{TR} is approximately the time required to bring the source of the FET that is directly connected to the output node either to V_{DD} or to ground. From the physical meanings, T_D and $T_{R/F}$ are generally of similar magnitude. Then the crisp waveform is an idealization that is not consistent with the circuit theory. This inconsistency is the basic reason that new, quantum-mechanical concepts must be used to understand circuit-level problems.

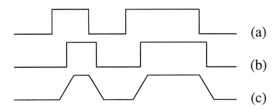

Figure 1.08.1 Idealized and real node waveforms

A logic circuit falls between the two distinct operational regimes: (1) At the low clock frequency, the sum of the rise-fall times of a pulse is negligible compared with the clock period; The node waveforms are like those shown in Figs.1.08.1(a) or (b). In the zero clock frequency limit, the logic circuit's throughput increases proportionally with the clock frequency; (2) At the high clock frequency the pulse rise-fall times are still less than the clock period, but they are the same order of magnitude. In the limit of high-speed operation, some or none of the nodes of the logic chain reach the steady Boolean-level during the clock period. In regimes (1) and (2) the boundary depends on the complexity of the logic circuit. Even at the low clock frequencies the special *critical* path may operate in regime (2), and this regime always determines the circuit's performance limit. Conversely, some short logic chains of a high-speed circuit may work in regime (1). This issue is related to how the logic circuit is designed.

One important point is that waveform crispness and delay certainly can be traded by the choice of the circuit designer. Lightly loaded inverters are often used to regenerate a crisply switching waveform from a badly corrupted waveform, as shown in Fig.1.08.2. The corrupted waveform of node M is reshaped to the crisply switching waveform at node Z

Classical Gate Field

by the inverter. The problem associated with this designer's choice is that the inverter's switching threshold voltage is uncertain. Depending on whether the inverter V_{GSW} is high or low, the regenerated waveform has a different pulse width. Then the essential attribute of the signal is altered, and the delay definition of the NAND gate does not make sense any more. The measured delay is the delay of the combined two stages or of the noninverting AND gate. Again, at the low-speed regime (1), where a lot of timing margin exists, the logic circuit can be designed including buffers in the strategic location where the waveform is corrupted. Then the uncertainty in the gate-chain delay becomes somewhat less. Such a freedom never exists in ultrahigh-speed circuit design. This is the practical problem encountered by the high-speed circuit designers.

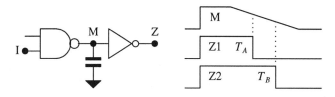

Figure 1.08.2 The effects of a waveform reshaping buffer

The effects of the rise-fall times become generally more significant in the higher clock frequencies, if the processing technology used to fabricate the circuit remains the same. The node voltage does not stay in the range where the Boolean-level can be determined unambiguously for a significant fraction of the clock period. At the present time, when the higher and higher clock frequency is demanded and technology upgrading occurs less frequently than it did in the 1980s, understanding the consequences of the less and less ideal node waveform to the circuit operation is quite important. In Sections 1.09 to 1.11 I discuss how the effects of the rise-fall times affect the circuit, both theoretically and practically.

The problem we discuss may be considered as caused by a mismatch between logic-circuit theory and electronic-circuit theory. The mismatch is exposed from the following observation. The idealized node waveforms shown in Figs.1.08.1(a) and (b) have the attributes of a rigid body in mechanics, whose location can be specified accurately. The real node waveform shown in Fig.1.08.1(c) has the circuit-parameter-dependent rise-fall times, and it has has the attribute of a soft, deformable body. The location of such an deformable object cannot be specified better than the overall size of the object: this is the same problem as being able to determine the position in the air of an aircraft, but not of a balloon. What is interesting is that this intuitive difference brings quantum-mechanical concepts into the theory, but the quantum mechanics in this problem has a certain flavor of classical mechanics. Mathematically, the quantum-mechanical flavor originates from the following: instead of using a time-dependent coordinate to describe a mass point, we need a time-dependent function to describe a deformable object.

This problem can be further explained as follows. Digital-circuit theory has a need similar to that of classical mechanics: the location and velocity of the information-carrying object must be determined accurately. If the need is satisfied, a theory closely in parallel to classical mechanics can be built, and this is logic-circuit theory. In logic-circuit theory, the idealized crisp node-voltage waveform can be assumed without causing consistency problems into the rest of the theory. Logic-circuit theory is a consistent classical-mechanics like theory. If the need is not satisfied, the theory must have some features of quantum mechanics, and the electronic theory of digital circuit is such a theory. In the theory the node-voltage waveform must be properly represented.

In this chapter we discuss the theoretical issues without going into the quantum-mechanical complications. To determine the location of an object that has a spatial size, we need to set a reference mark on the object that is much smaller than the object itself. How can a mark be set on the moving stepfunction of voltage in a digital circuit? The most obvious way is to set a threshold voltage of the Boolean-level, as we have done. Can this be set uniquely and consistently? This is the question we study first.

1.09 Logic Threshold Voltage and Gate Delay Time

Gate delay time is conventionally defined as the interval between the time when the gate's input voltage reached the logic-threshold voltage and the time when its output voltage reaches the same logic-threshold voltage. This definition has been applied not only to a single gate but also to a cascaded chain of gates in both inverting and noninverting versions. To be consistent with this definition, the *logic − threshold voltage* is defined as follows. A logic node has a Boolean-level HIGH if the node voltage is higher than the logic-threshold voltage and has a Boolean-level LOW if the node voltage is lower than that. Then the delay time of a gate is the time required to create the Boolean-level answer at the output, after the Boolean-level at the input is established at the threshold level. This is a time domain observation. In a space domain, the time when a signal front arrives at a logic node can be defined from a similar observation. The velocity of the signal in the unit of the node index (the location of node in a sequential integer number) per second is defined. This definition, which I call the *classical delay − time definition*, is necessary more for practical convenience than for basic theoretical reasons, and it contains many difficulties.

Classical Gate Field

The first problem is the choice of the logic-threshold voltage itself. Let us look at the problem from the broadest angle: the threshold voltage divides the range of the node voltage swing into a region where the Boolean-level is HIGH, and a region where the Boolean-level is LOW. A single, continuous dynamic range of the node-voltage swing is arbitrarily divided into two regions having different Boolean-levels. Intuitively, this is not convincing since there is no reason based on the physics of the circuit, and this point becomes most clear if we consider where to set a logic-threshold voltage acceptable for all the gates.

In CMOS, the logic-threshold voltage is conventionally set at $V_{DD}/2$ (Shoji, 1987). This setting intuitively appears rational, since a CMOS inverter can be symmetrical with respect to pullup and pulldown by selecting the device parameters. Yet this symmetry demands too much restriction and idealization to CMOS static circuits in general: it is valid only for an inverter and not for any other multiinput gate. The combinational gates are, by their construction, pullup-pulldown asymmetrical. For a logic circuit that contains many different types of gates that are built using practically any size FETs, to choose a single logic-threshold voltage by definition is too restrictive and unnatural, even if we consider that the gates are used in quite restricted operating conditions. In spite of the problem, the definition is used conventionally to compile the design information.

The practical advantage of this definition is that if certain additional requirements (such as the gate pullup and pulldown capability) are within limited range, the sum of the delay times of the gates in a logic chain is a reasonable approximation of the delay time of the entire gate chain, which is determined by the same delay definition. This convenience is indispensable in commercial chip design, but it is not necessarily real or accurate, especially in ultrahigh-speed circuits. Since this approximation assumes that the gates are perfectly enabled or that every input (other than that connecting the gates into a gate chain) is all at the appropriate Boolean HIGH or LOW level voltage, the effects of simultaneous or near simultaneous switching of more than one input are excluded. If many such logic chains make a complex web of logic gates and its input data are sourced by two or more latches that are clocked by the circuit's clock, simultaneous switching can never be avoided. It is too arbitrary to identify only one signal delivered from a specific latch as the cause of the transient induced in the logic chain (Section 2.13): the cause of the transient must be the latch clock, which lets all the latches deliver the signal simultaneously. Neglect of the simultaneous switching effect may create gross error in the delay-time estimate to the optimistic side. This effect establishes an important requirement for the CAD tools used in design verification of ultrahigh-speed circuits, as I discuss in Section 3.15. Nevertheless, a gate-level delay simulator uses the delay-time sum approximation all the time. This approximation is very curious and is almost empirical in nature: it is quite impossible to derive it theoretically or to determine the degree of approximation on a theoretical basis.

Let us study, by examples, how well the classical delay definition works and how reasonable it is. We consider first a single-input, single-output inverting logic gate. A multiinput-logic gate can be a single-input logic gate, if the inputs (other than the particular input driven by the signal source) are set at the Boolean-level that *enables* the gate. Then the gate becomes logically equivalent to an inverter. An enabled logic gate as an equivalent inverter has its own degree of asymmetry with respect to pullup and pulldown operations. Let us consider how much asymmetry exists in conventional CMOS gates. A convenient single measure of the asymmetry is the gate's switching-theshold voltage V_{GSW}.

An enabled logic gate is in the most active state if the input and output voltages are equal. At this voltage small variations of the input voltage are transmitted to the output with voltage gain. This voltage, the gate-switching threshold voltage V_{GSW}, is determined by connecting the particular input and output of an inverting and enabled gate in the powered-up condition. The inverting gate is in a negative feedback loop, and if the gate is the one-stage inverting type, the common node voltage settles stably at the V_{GSW}. Table 1.09.1 gives the gate-switching threshold voltages of simple CMOS gates of 0.35 micron CMOS, built from the minimum-size NFET and the second minimum-size PFET. The operating conditions are typical conditions V_{DD} at 3.3 volts, 85 degrees centigrade, and typical processing conditions. In a multiinput gate, more than one V_{GSW} exist, since there are choices of which input terminal is connected to the output terminal. The other terminals are all connected to either the HIGH (V_{DD}) or the LOW (V_{SS}) logic-level voltages that enable the gate. Input 1 drives the FET closest to either V_{DD} or V_{SS} bus, and inputs 2, 3 . . . , are successively away from either power rail. If all the inputs are connected together, the gate is in the simultaneous switching mode. The V_{GSW} for the simultaneous switching condition is also shown. They are quite different from the single-input V_{GSW}. As we observe, just this much definition requires many specifications of the operating condition. Yet each V_{GSW} is for only one of many ways to operate the gate.

From Table 1.09.1 we observe that V_{GSW} can be anywhere between V_{THN} and $V_{DD} - V_{THP}$, that is, the gate voltages where the NFET and the PFET turn off, respectively. Since PFET has higher conduction-threshold voltage and less conductivity, the V_{GSW} are more frequently below $V_{DD}/2$, but this is not crucial in the following argument. These data show that CMOS multiinput gates are generally quite pullup-pulldowm asymmetrical. As is observed from the NAND4 and NOR4 data most clearly, V_{GSW} depends on the modes of operation and on the selection of the inputs. The important point is that the logic gate must produce the correct Boolean-level at the output irrespective of the mode of operation. From this observation, selection of the logic-threshold voltage based on the pullup-pulldown symmetry has only weak and unconvincing theoretical basis.

Table 1.09.1: Switching threshold voltage					
Gate type	Input 1	Input 2	Input 3	Input 4	Simul.
INV	1.5260	*	*	*	1.5260
NAND2	1.5390	1.6425	*	*	1.9047
NAND3	1.5547	1.6513	1.7263	*	2.0864
NAND4	1.5838	1.6610	1.7327	1.7904	2.2004
NOR2	1.4982	1.3595	*	*	1.1016
NOR3	1.4502	1.3431	1.2530	*	0.9198
NOR4	1.3783	1.3173	1.2418	1.1779	0.8085

We note the another problem: the classical delay-time definition is based on the gate's DC operation. In a system's operational environment, a logic gate practically never makes a Boolean-level determination in the static mode. The timing of the signal of different inputs is not certain: the voltage that should have enabled the gate earlier may arrive almost simultaneously. The gate inputs are driven hard and rapidly, and it is conditioned to make the decision quickly in the dynamic switching mode of operation. Then the gate's dynamic switching threshold becomes the issue. Figure 1.09.1 shows the dynamic switching characteristics of a 0.35 micron CMOS inverter, NAND2, and NAND3 gates, measured at the input voltage slew rate of 1 volt per nanosecond.

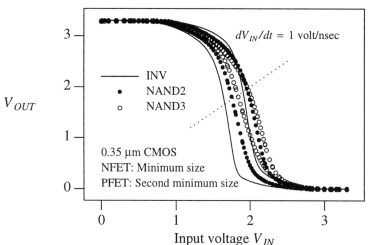

Figure 1.09.1 The input-output characteristics of the inverter, NAND2, and NAND3 gates at high speed

The input pullup transitions are on the left side, and the input pulldown transitions are on the right side. Since the input-output voltage relationships are different for the two switching polarities, no single gate-switching threshold voltage can be defined in the dynamic operation. The input-output voltage relationship depends on the slew rate of the input voltage. This arbitrariness meshes awkwardly with the classical delay-time definition.

Based on the observations made in this section, there is no convincing reason why the logic-threshold voltage should be set at $V_{DD}/2$ or at any particular level between V_{TH} and $V_{DD} - V_{THP}$. Isn't it true that any voltage in the range can be the logic-threshold voltage, with equal reasons for and against setting voltage there. There is a reason, however, not to set the threshold voltage outside the range: the gate output voltage is guaranteed to be V_{DD} or V_{SS}, at least in the static operation (we neglect FET leakage currents). How, then, shall we proceed? This is the question answered adequately only by the quantum-mechanical model of logic circuit, which is discussed in Chapter 2. Before discussing the issue, we need to investigate the problems of the classical delay-time definition. Other than the choice of the logic-threshold voltage, there are many strange consequences that follow from the definition.

1.10 Nonmonotonous Node-Switching Voltage Waveforms

The first problem is described as follows. If a node-voltage waveform is not monotonous, delay-time definition by using the logic threshold voltage may become absolutely impossible. An example is shown in Fig 1.10.1, where the input waveform is shown by the dashed curve, and the output waveform is shown by the solid curve.

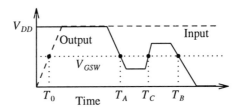

Figure 1.10.1 Nonmonotonous node-voltage waveform and delay definition

If the gate or the gate-chain output waveform is nonmonotonous, as is schematically shown in the figure, it is unclear, whether the chain's delay time is $T_A - T_0$ or $T_B - T_0$. This is the case if the gate or the gate chain is known, a priori, to be inverting. If even that is not known, $T_C - T_0$ can be identified as a possible delay time as well. This difficulty creates serious problems in the logic-chain delay definition, and it suggests that there is a fundamental difficulty in the concept of gate delay time as a

Classical Gate Field

well-defined parameter that can be determined purely by mathematical operation, such as searching for an intersection of a curve and a straight line (for instance, by using a computer). The classical delay-time definition and the quantum-mechanical definition, given in the next chapter are qualitatively different ways of handling this problem.

Here we considered the problems of nonmonotonous node waveforms from their basic mathematical definition first. In an intuitive model of a digital circuit, it is considered possible to describe the circuit operation by propagation of a single upgoing or downgoing stepfunction, from beginning to end. Practically, this is too strong a restriction to digital-circuit operation: the idealization is possible only when the logic circuit does not have a *recombinant* fanout that combines the signal stepfunction edges having different polarity, both originating from the single source latch clock (Section 2.16). If a NAND or NOR gate combines signals that have opposite transition polarities that arrive at different times, inevitably there is a chance of generating an isolated pulse, referenced to either a Boolean LOW or a Boolean HIGH level. The isolated pulse is referred either to a flat logic level or a stepfunction. Let us first give a simple example of this phenomena, which is called a *hazard* (McCluskey, 1986; Mano, 1991; Nagle, Carrol and Irwin, 1975).

Figure 1.10.2(a) shows a digital circuit that creates an isolated pulse or its smoothed-out version, a nonmonotonous node waveform. All the signals A, B, and C, originate from a bank of clocked latches. The latches deliver the stepfunctions synchronously. The signals travel through the combinational circuits to the inputs of the NAND and NOR gates, A, B, and C, with different delay times for each. Suppose that node A makes a LOW to HIGH transition and node B makes a HIGH to LOW transition. If the delay time to node A is shorter than the delay time to node B, NAND gate output makes a HIGH to LOW transition at the arrival of signal A and then makes a LOW to HIGH transition at signal B. An isolated pulse is generated at node X.

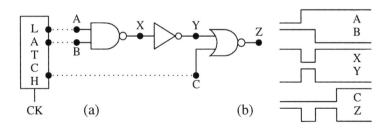

Figure 1.10.2 Generation of
a nonmonotonous node-voltage waveform by a recombinant fanout

Suppose that the node X signal is relayed to the downstream by the buffer and drives the NOR gate. If signal C, which has still longer delay time than signals A and B, is combined to signal Y at the NOR gate, the output of the NOR gate first makes a HIGH to LOW transition, followed by a LOW to HIGH transition on arrival of the isolated pulse from node X. Then the main transition by signal C takes place. A three-step transition waveform shown in Fig.1.10.2(b) is generated. The isolated pulse at node X is called a *static hazard*, and the orphan pulse at node Y is called a *dynamic hazard*. Obviously, each transition edge is caused by a single cause a stepfunction originating from the bank of the latches. The transitions do not line up. Then, if the path delay times are not zero, there is always a possibility of generating the isolated pulse.

As is seen from this example, for every new extra transition edge to form, there must be a cause, which is a stepfunction signal originating from the clocked latch. The latch clock is the signal that is identified as the single cause of the event in the clock cycle. Then the number of combined signals originating from a single clock pulse minus one is the maximum number of extra edges. It is impossible to exclude the non-monotonous node waveform issue from digital circuit theory.

In Section 1.08 we discussed the possibility of using a fast inverter or buffer to reduce the signal to generate a crisp switching waveform for better delay-time definition. If the output-node waveform including isolated pulse, such as that shown in Fig.1.10.2(b) is generated in the circuit, a fast-switching buffer added to the end of the logic gate chain makes the isolated pulse even more clearly observable. This means that the method aggravates the hazard. To remove the isolated pulse it is necessary to deteriorate the high-speed response. The two objectives are contradictory. From this viewpoint, the most ideal logic gate that maintains single-stepfunction propagating all across the logic circuit must have a properly compromised, not too fast, not too slow response. The rise-fall times of a gate must be short enough to secure clear classical delay-time definition but long enough to erase the extra pulse or glitch. This is too much to hope for. The model based on single-stepfunction propagation is fundamentally inconsistent. Classical delay-time measurement indeed shows too many basic problems on close examination. Delay time measured on such a shaky ground approximates, at best, with the unknown validity.

Even if the classical logic-circuit model is adhered to, the circuit operation can never be described by only one stepfunction front arriving at the destination. There can be more than one stepfunction front in general, and the waveform from the combinational logic circuit must be *interpreted* at the destination (Shoji, 1992). Classical digital-circuit theory requires intelligent observation of the node waveforms to extract the Boolean-level. The final result of digital signal processing can be either: (1) An even number of logic-level transitions that maintain the output

logic level the same, or
(2) An odd number of logic-level transitions that inverts the output logic level.

The problem is, *logic – level transition* comes with node waveform deformation, which may have an extra pulse, a nonmonotonous waveform, or a plateau in the middle of the transition step. One way to look at the digital-circuit operation is that it is the process of creating and erasing glitches. This may create a lot of ambiguity in the classical definition of delay time. If signal delay time is defined classically, the nonmonotonous waveform requires intelligent interpretation at the destination, and that is the crucial operation of a digital circuit. Since a combinational logic circuit generates correct output in the limit an infinitely long time after the input logic level settles, the intelligence required to interpret the waveform to extract the Boolean-level and the waiting time can be traded. Classical logic-circuit theory assumes the limit of long waiting time and hides the requirement of intelligence. Yet hiding the requirement creates a problem if delay time becomes the issue. The problem is worse in high-speed circuits.

Nonmonotonous waveforms create significant delay uncertainty in a short chain of logic gates. The uncertainty can be less significant in a long chain of logic gates. Long or short, the problem exists. We need an intelligent observer of a digital signal. A latch has an elementary intelligence, as we see in Section 2.15. If the logic chain is driven by latches, and the chain outputs are captured by latches, the uncertainty decreases since a latch has a certain capability to interpret the waveform. A logic circuit may be viewed as a combination of the mechanical generation of output waveform by the combinational logic circuit and an intelligent observer, the output latch, that interprets and reshapes the signal. This point is fundamental to the quantum-mechanical delay definition discussed in Chapter 2.

1.11 The Strange Consequences of the Classical Delay-Time Definition

According to the theory underlying this book, digital information is carried by an identifiable object, and the two types of object, upgoing and downgoing stepfunction voltage wavefront, propagate through a digital circuit. The Boolean-level of the stepfunction at the driving point is the signal level, which propagates through the logic circuit. This interpretation is rational because obviously there is a flow of information in the digital circuit and the location of the information carrier is always identifiable. In such cases, the information-carrying object appears to be a particle. If the motion of the information carrier can be explained by the motion of the particle, the model is simple and complete, but this is not always the case. The problem of this interpretation can be demonstrated by identifying the cases where such a simple interpretation leads to inconsistencies and difficulties. In the last section we saw one example the difficulties associated with a nonmonotonous node waveform. We expand our search to other awkward cases. This is a prerequisite to exploring the foundation of the digital-circuit model.

There is an issue that requires attention in the definition of classical delay time. Since the gates are arranged in the order of signal propagation, each gate's location is identified by a single integer number in increasing order 1, 2, 3, . . . , N, as is shown in Fig.1.11.1. The information-carrying signal should travel in the increasing order of the index. We note that the structure of the circuit shown in the circuit diagram is not always clear enough for us to assign the number from one end to the other. If the delay time of a gate at index i, T_i, is determined (delay time from node $i-1$ to i), the velocity of the signal front there is determined as $1/T_i$ (stages per unit time). This definition appears obvious, but it has a few hidden awkward assumptions. The definition assumes that the signal front passes a logic node only once. In an infinitely long gate chain excited at the end as shown in Fig.1.11.1, the assumption is satisfied, but it is not satisfied in a gate chain that makes a loop, especially a short loop. If this assumption is rephrased, the spread of the stepfunction must be confined within a small number of stages, only one or two in a long path. If the gate chain makes a loop, an origin of the node index can be set only if the information-carrying signal front passes the origin at definite and mutually well-separated time points. The index setting makes sense only if the condition is satisfied. If the gates make an N-member loop (node $N+1$ and node 1 are the same node), the change of the node voltage at node index i should not influence the node voltage at node index i by going through the origin of the index, node 1. If this does not happen, the chain can be cut at node 1 and the index can be assigned to the nodes. If the node i voltage affects itself by going through the origin, node 1, and completes the circle, it is impossible to cut the loop open at node 1, and therefore the velocity of the stepfunction cannot be defined. Then it appears to the observer as if all the node voltages change simultaneously.

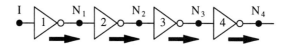

Figure 1.11.1 Cascaded inverter chain

The loop nodes interact strongly if the number of the gates making up the loop is small. An inverting one-gate chain (a single gate whose input and output are connected together) settles stably at a DC bias point V_{GSW} (Section 1.09). A two-gate chain is the circuit used as a binary latch. In the latch circuit the two nodes interact so strongly that the chain cannot be broken open without affecting the transient in it. If the two nodes interact, the transition is not localized at any one node, and the two-node voltages change simultaneously. Then the location of the excitation can never be established. A latch consisting of two inverter stages is a circuit in which the excitation is spread over the loops. A binary latch is indeed a very special circuit, and this is the crucial point of the quantum-mechanical definition of the gate delay time, which is introduced in Chapter 2. The two nodes of a latch may be considered to be a node having the same signal, except that the one is the complement of the other. The time references of the two nodes are forced to come close together. This is the reason that a cross-coupled pair of inverters is used to reduce the skew of a clock and its inversion. If the discussion of the quantum-mechanical delay definition (see Sections 2.09, 2.14 and 2.15) is examined from the viewpoint that stresses the special character of a latch made of a two-inverter closed loop, it is not so bizzare as it might be felt intuitively. A three-inverter closed loop makes the simplest ringoscillator. In this circuit, the location of excitation is moderately well defined, as is discussed in Section 3.02.

Another strange consequence of classical Boolean-level determination is observed from the latch circuit shown in Fig.1.11.2. Let the latch-circuit nodes N_1 and N_2 be clamped at the voltages V_1 and V_2, respectively. As the clamp is released, the node voltages change, and the latch settles at one of the two states in one N_1 is HIGH and N_2 is LOW, and in the other N_1 is LOW and N_2 is HIGH.

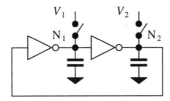

Figure 1.11.2 Latch-circuit delay time

Since the inverters are connected as a closed loop, the information-carrying particle must have circulated the loop and finally settled at one position or the other. Suppose that the inverters have the same V_{GSW}. If V_1 and V_2 are different but both higher than V_{GSW}, both nodes must have had the HIGH Boolean-level at the beginning. One of the HIGHs changed to the LOW. If V_1 and V_2 are both lower than V_{GSW}, one of the LOWs changed to the HIGH. In this case we consider that a particle is created by the latch. In a digital circuit an information-carrying particle can be created and annihilated. This is not a property of a classical particle or a mass point: they are more appropriately called *pseudoparticles*, such as phonons in a crystal lattice, and they are the subject of quantum mechanics.

Let us go back to the problems of an open-gate field. Even if the node-voltage waveform is monotonous, the following strange problem appears. Let us consider the cascaded inverter chain shown in Fig.1.11.1. Let the switching-threshold voltage of inverter 1 be quite low and that of inverter 2 be quite high. Suppose that the input voltage of node I increases slowly from 0 to V_{DD}, as is shown in Fig.1.11.2. Let the logic-threshold voltage be V_{L1}. We find the arrival of the signal front at node I at time T_I. Similarly, the arrival time of the signal at node N_1 is T_1, and the arrival time at node N_2 is T_2. From Fig.1.11.3, $T_2 < T_1 < T_I$. This means that digital front arrives at N_2 first, then at N_1, and thereafter, the source node I. The front is moving backward in the inverter chain. This is obviously an absurd conclusion. Yet there is no error in the reasoning that leads to this conclusion. This conclusion depends on the selection of the logic-threshold voltage at V_{L1}. If it is chosen at V_{L2}, the information-carrying front appears to move from N_1 to N_2 and then back to I. If the threshold voltage is chosen at V_{L3}, the front appears to move from I to N_2 and to N_1. Obviously, if the location of the digital signal in a logic chain is determined by the classical criteria by setting the logic-threshold voltage, many conclusions, which are mutually contradictory, are reached.

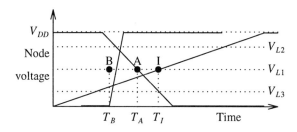

Figure 1.11.3 The node-voltage waveforms of an inverter chain

The delay time of a dynamic (precharged) CMOS gate is confusing in several respects. Figure 1.11.4 shows a cascaded Domino CMOS OR gate chain (Shoji, 1987). The NFETs surrounded by the dotted box are called *ground switches*. Ground switches can be eliminated except for the first stage of the chain, where input signals are introduced. Before the logic operation, clock $\overline{CK}$ is pulled down to the LOW logic level. The precharge PFETs pull the dynamic nodes up to the HIGH logic-level voltage, and the buffered outputs, B, C, and D, are pulled down. The delay time to pull the nodes down is the interval between the time when the clock goes down to the time when nodes B, C, and D go down. The delay time has two confusing features: (1) The delay is independent of

Classical Gate Field 63

the length of the cascaded Domino CMOS gate chain, and (2) the delay time has apparently nothing to do with the signals that drive the chain.

As for feature (1) we note that the circuit is reconfigured during the precharge operation. In the precharge circuit configuration all the gates are precharged in parallel, and the delay time is that of a single Domino CMOS gate (a composite of the NFET-based dynamic logic circuit and the inverter). If the ground switch is provided only to the first stage, the precharge delay time extends to the delay time over several stages of the Domino CMOS gates, so problem (1) is modified accordingly. As for (2) we need to observe the operation during the discharge (evaluation) phase.

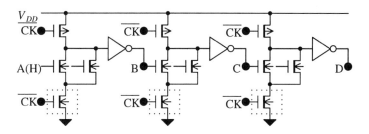

Figure 1.11.4 A Domino CMOS delay definition

During the precharge phase, the input signals to the Domino CMOS chain is stabilized to the final level, and they are held unchanged during the subsequent evaluation phase. After the precharge is completed, clock $\overline{CK}$ makes a LOW to HIGH transition. The precharge PFETs are turned off, and the ground switches are turned on. If any of the NFETs of the precharged dynamic NOR gates has been turned on by a HIGH input signal level, the precharged node is pulled down, and the buffered output node is pulled up. Then the NFET of the next-stage Domino CMOS gate is turned on, and so on. The wave of discharge processes propagates down the logic chain, which resembles falling dominoes. Let the starting point be A(H) of Fig.1.11.4. The delay time of the logic operation is the propagation delay time of the domino wave, which starts with the clock transition. Since the input data must have been stabilized before, the delay is referred to the clock, which is the cause. If none of the input node is HIGH, however, the state of the output node does not change. Does this mean that in this case the delay time of the chain is zero? This is obviously an absurd conclusion.

The signal that initiates discharge is the clock $\overline{CK}$ LOW to HIGH transition and not the transition of the input signal. The confusion, as well as problem (2) in the precharge phase, can be resolved naturally by changing the view of the digital signal. A digital signal is a time-independent information, and it is the clock that creates the propagating transition edge. The delay time is always referenced to the clock, if the origin of the signal is tracked down to the source. We came across another quite fundamental issue: the circuit's clock is essentially in the definition of the logic gate's delay time. This viewpoint is further collaborated in Sections 2.14 and 2.15, where the quantum-mechanical delay measurement is initiated by the clock of the test equipment.

1.12 The Phase Transition of the Gate Field

The structure of the discrete gate-field making a regular lattice structure resembles a lattice of a crystal. One of the most remarkable thermodynamical properties of crystal is that it goes through a phase transition. If the temperature is raised, a crystal melts and loses its regular structure, and an associated thermodynamic effect, like absorption of the latent heat, is observed at the transition. Phase transition is a fascinating thermodynamic phenomenon, in that it makes a qualitative decision about state based on quantitative field variables like temperature and pressure. It thus resembles a quantum-mechnical transition but occurs in a macroscopic scale. Therefore, a phase transition can be examined closely by classical physics. The gate-field displays a similar phase transition.

To demonstrate the simplest example of a gate-field phase transition, let us consider the one-dimensional gate-field shown in Fig.1.12.1. This is a parallel combination of an RC chain circuit and an inverter chain, sharing the node capacitance C. The inverters try to force the node voltages to alternately HIGH and LOW digital levels, and the resistors try to maintain them all equal, somewhere between the two Boolean-levels. The gate-field has two physically distinct extrema states: in one the FETs making the inverter are much less conductive than resistance R, and in the other the FETs are much more conductive than the resistance. In the first state the node-voltage profile is more or less uniform or monotonous from one node to the next node, and in the second it switches from V_{DD} to V_{SS} from one node to the next.

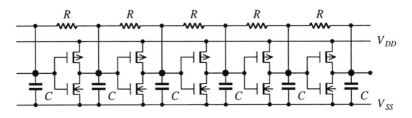

Figure 1.12.1 A paralleled inverter-RC line gate-field

Classical Gate Field

Suppose that the inverters are pullup-pulldown symmetrical and that the FETs have transconductance G_m and the threshold voltage V_{TH}. At what value of R does the phase transition take place? Let the voltage of node n be V_n. It satisfies a set of circuit equations

$$C(dV_n/dt) = [(V_{n+1}+V_{n-1}-2V_n)/R]+I_P-I_N$$

where I_P and I_N are the PFET and the NFET currents, respectively. Let the FET currents be given by the collapsable current generator model (Shoji, 1992) and let the drain to source voltages not be zero.

$$I_P = G_m(V_{DD}-V_{n-1}-V_{TH}) \quad \text{and} \quad I_N = G_m(V_{n-1}-V_{TH})$$

The assumption of a nonzero drain to source voltage is justifiable in the transition starting from the uniform state and ending at the alternating state. The steady-state solution of the equation is

$$V_{n+1}+V_{n-1}-2V_n+G_mR(V_{DD}-2V_{n-1}) = 0$$

and $V_n = V_{DD}/2$ for all n is the steady-state solution. Let the small variation around the steady state be written as

$$V_n = (V_{DD}/2)+v_n$$

Then v_n satisfies the equation

$$dv_n/d\theta = (v_{n+1}+v_{n-1}-2v_n)-Gv_{n-1}$$

where $G = 2G_mR$, and where $\theta = t/RC$ is the normalized time. We search for a solution having the form

$$v_n = \alpha^n \exp(S\theta)$$

By substitution we find that S and α satisfy

$$S(\alpha) = \alpha-2-[(G-1)/\alpha]$$

For the inverter to win over the resistor, $\alpha = -1$ is required. For this α, if we set S to zero, the critical parametric boundary between growth and decay, we have $G = 4$. Then we expect that at $g_mR = 2$, the phase transition takes place. Figure 1.12.2 shows the results of a numerical analysis. The length of the gate-field is 20, and the parameter values are given in the figure. The voltage of the first node is set at $V_0 = 0.3 \times V_{DD}$. For smaller R the field is uniform at $V_n = V_{DD}/2$. For larger R the node voltages approach a state in which V_{DD} and V_{SS} alternate. The gate-field goes through a phase transition if R is changed across the critical value.

We observed the phase transition between the uniform and the alternating gate-field-voltage profile. The uniform phase consumes power: currents flow from V_{DD} to PFET to NFET to ground at each inverter location. The alternating phase does not consume that much power, since either PFET or NFET at any location is turned off and R is high. A phase transition of this type (one phase consuming more power than the other) has been observed in bulk negative resistance material like N-type gallium arsenide (Ridley, 1963).

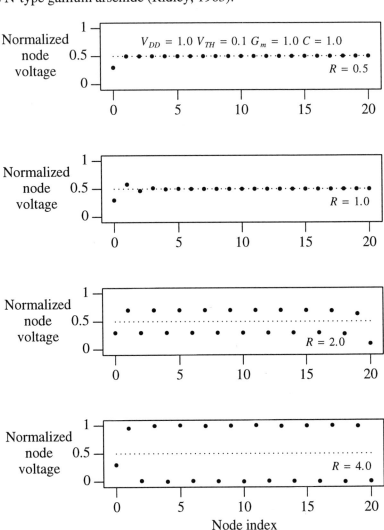

Figure 1.12.2 Phase transition
in a paralleled inverter-RC chain

The alternating phase consists of two different variations: in one the even-numbered nodes are HIGH and the odd-numbered nodes are LOW, and in the other, vice versa. The two phases are the subphases of the alternating phase, and they are thermodynamically equivalent. Such a phenomenon is observed in ferromagnetic material. If a ferromagnetic crystal is held at high temperature, it is in the uniform paramagnetic phase. If the temperature is reduced below the Curie point, the crystal goes through a transition to the ferromagnetic phase, which has several subphases (Carey and Issac). In each of the subphases the magnetization

Classical Gate Field

is aligned to one of several easy magnetization axes of the crystal. The two equivalent subphases of the alternating phase of the gate-field are equivalent to the equivalent subphases of ferromagnetism.

The idea that a gate-field goes through a phase transition may be unconventional, but that is primarily because the phenomena that the field support and the structure of the field have been considered separately. The two phases are *functionally* distinct but structurally the same, and the difference is caused by the resistance parameter value. What goes through the phase transition is *not* the circuit hardware but the excitation in it. The idea is not strange but is rational and interesting.

1.13 The Miller Effect in the Gate Field

We consider a gate-field structure similar to that shown in Fig.1.12.1, but having components other than resistors connecting the nodes. If the components are Miller-effect capacitors C_m as shown in Fig.1.13.1, numerical analysis becomes difficult because the set of the circuit equations cannot be solved for the derivatives of the node voltages dV_i/dt ($i > 0$) in a simple form. The solution requires a matrix inversion. This is quite cumbersome. Therefore, we study an example of the reasonably long, and still mathematically manageable six-inverter Miller-effect gate-field of Fig.1.13.1 Let the PFET pullup current and the NFET pulldown current of the n-th stage be $I_P(V_{DD},V_{n-1},V_n)$ and $I_N(V_{n-1},V_n)$, respectively. The difference in the two currents charges or discharges node n capacitance C, and the current is written as

$$\Delta I_n = I_P(V_{DD},V_{n-1},V_n) - I_N(V_{n-1},V_n)$$

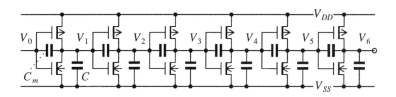

Figure 1.13.1 The effects of interstage coupling capacitance

Then the circuit equations are written as

$$C\frac{dV_i}{dt} = \Delta I_i + C_m \frac{d}{dt}(V_{i-1}+V_{i+1}-2V_i) \quad (1 \le i \le 5)$$

and

$$C\frac{dV_6}{dt} = \Delta I_6 + C_m \frac{d}{dt}(V_5 - V_6)$$

The last equation is different from the others, since the seventh stage never exists.

If we set $dV_i/dt = X_i$, a set of simultaneous equations for the derivatives are written as

$$(C+2C_m)X_1 - C_m X_2 = \Delta I_1 + C_m X_0$$
$$-C_m X_1 + (C+2C_m)X_2 - C_m X_3 = \Delta I_2$$
$$-C_m X_2 + (C+2C_m)X_3 - C_m X_4 = \Delta I_3$$
$$-C_m X_3 + (C+2C_m)X_4 - C_m X_5 = \Delta I_4$$
$$-C_m X_4 + (C+2C_m)X_5 - C_m X_6 = \Delta I_5$$
$$-C_m X_5 + (C+C_m)X_6 = \Delta I_6$$

where the first equation has the $C_m X_0$ term on the right side of the equal sign because it is determined from the given input node waveform.

From the set of equations, X_1 to X_5 are solved in terms of X_6 by using the recursive formula as

$$X_5 = (1/C_m)[(C+C_m)X_6 - \Delta I_6]$$
$$X_4 = (1/C_m)[(C+2C_m)X_5 - C_m X_6 - \Delta I_5]$$
$$X_3 = (1/C_m)[(C+2C_m)X_4 - C_m X_5 - \Delta I_4]$$
$$X_2 = (1/C_m)[(C+2C_m)X_3 - C_m X_4 - \Delta I_3]$$
$$X_1 = (1/C_m)[(C+2C_m)X_2 - C_m X_3 - \Delta I_2]$$

To express the result of the algebra conveniently we define a set of polynomials:

$$Q_0 = C^6 + 11 C_m C^5 + 45 C_m^2 C^4 + 84 C_m^3 C^3$$
$$+ 70 C_m^4 C^2 + 21 C_m^5 C + C_m^6$$
$$Q_1 = 1$$
$$Q_2 = C + 2C_m$$
$$Q_3 = C^2 + 4C_m C + 3C_m^2$$
$$Q_4 = C^3 + 6C_m C^2 + 10 C_m^2 C + 4 C_m^3$$
$$Q_5 = C^4 + 8C_m C^3 + 21 C_m^2 C^2 + 20 C_m^3 C + 5 C_m^4$$
$$Q_6 = C^5 + 10 C_m C^4 + 36 C_m^2 C^3 + 56 C_m^3 C^2 + 35 C_m^4 C + 6 C_m^5$$

By using Q_0 through Q_6, X_6 is solved as

$$X_6 = (C_m^5/Q_0)[C_m X_0 + \sum_{i=1}^{6} (Q_i/C_m^{i-1})\Delta I_i]$$

Using the solutions of the derivatives, the six-member Miller-effect gate-field of Fig.1.13.1 can be numerically analyzed. Two results for fast and slow input rise time t_1 are shown, respectively, in Figs.1.13.2(a) and (b). $V_0(t)$ is given by

$$V_0 = 0 \ (t < 0) \quad V_0 = V_{DD}(t/t_1) \ (0 < t < t_1)$$
$$V_0 = V_{DD} \ (t > t_1)$$

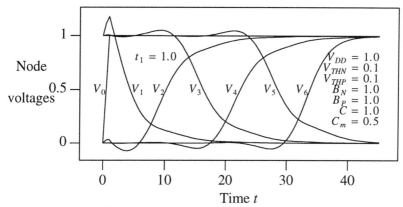

Figure 1.13.2(a) The node voltage waveforms of a Miller-effect inverter chain

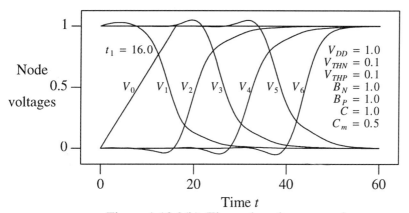

Figure 1.13.2(b) The node voltage waveforms of a Miller-effect inverter chain

The two sets of curves show a common feature, a *backkick* before going through either the upgoing or the downgoing principal transition. Capacitive coupling transmits an input voltage change to the output without delay. For a conducting FET to charge or discharge the node capacitor C, it takes time of the order of the product of C and the channel resistance of the FET. The backkick occurs before the charge and discharge process takes place. If node 0 voltage V_0 makes a LOW to HIGH transition, node 1 voltage makes a HIGH to LOW transition after the inverter delay time. If a Miller-capacitance C_m exists, it transmits upgoing V_0 to node 1 without delay, and node 1 voltage V_1 goes up temporarily above V_{DD} before the pulldown process begins. This backkick takes place at all the nodes of the gate-field, both in positive (nodes 1, 3, and 5) and in negative (nodes 2, 4, and 6) polarities. In the numerical analysis the FET currents are computed by allowing the effects of the temporary switch of the source and the drain identification of the FET. For node 1 the height of the backkick is determined by the input-voltage V_0 rise time. For the two sets of curves of Figs.1.13.2(a) and (b) the height of the backkick of node 1 decreases as the input-voltage rise time t_1 increases. In the nodes some distance away from the driving end, the height of the backkick is determined by the characteristic rise-fall times of the Miller-effect gate-field itself. The heights of the backkick for Figs.1.13.2(a) and (b) for V_3 through V_6 are about the same. The effective node capacitance that determines the delay and the rise-fall times is approximately $C + 4C_m$, since a single C_m works effectively as a capacitance of twice the nominal value because the two terminals are driven by antiphase signals. Furthermore there are two C_m one to the left and the other to the right sides.

Figures 1.13.2(a) and (b) show the cases where $C_m < C$, or the Miller capacitance is less than the node capacitance. If $C_m > C$, the backkick becomes more pronounced, and the distortion of the node waveform becomes noticeable, as is shown in Fig.1.13.3.

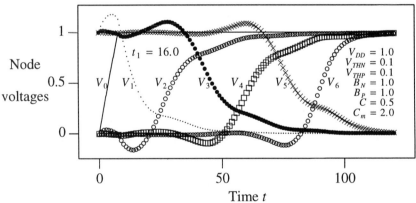

Figure 1.13.3 Node-voltage waveforms
of a Miller-effect inverter chain

Let us observe the HIGH to LOW transitions of V_1, V_3, and V_5. In the later phase of these pulldown transition, there is a characteristic *hump*, where the decrease of the node voltage becomes temporarily slow. In the *hump* region we observe the following: for the V_1 waveform the V_2, for the V_3 waveform the V_4, and for the V_5 waveform the V_6 waveform

Classical Gate Field

intersects in the hump region. The intersecting waveforms V_2, V_4, and V_6 at their respective intersections have a high rate of voltage increase. The increase of the output voltage of an inverter drives the input of the inverter up, thereby slowing the input-voltage change temporarily. The same interpretation holds for the humps of the upgoing waveforms V_2 and V_4 (V_6 is the end of the chain, which does not suffer from the Miller effect). There are other fine structures in the node waveform. For the V_4 waveform, the effects of capacitive coupling from decreasing V_3 at about $t = 45$, from increasing V_2 at about $t = 25$, and from decreasing V_1 at about $t = 5$ are also recognizable.

The current from the input of the n-th inverter to its output through the Miller-capacitance C_m is given by

$$I_{Mn} = C_m (d/dt)(V_{n-1} - V_n)$$

Figure 1.13.4 shows waveforms of I_{Mn} ($n = 2, 3, 4, 5,$ and 6). $I_{Mn} < 0$ if n is even (the inverter input pulls down and the output pulls up), and $I_{Mn} > 0$ if n is odd (vice versa). The current pulse width of I_{Mn} is wider than the inverter delay time, since the Miller current flows if either node voltage changes. The current waveform consists of two peaks one due to V_{n-1} change and the other due to V_n change. I_{M6} has a single peak, since there is no stage to the right. The maximum of the Miller current is estimated from the rise-fall time $t_{R/F}$ as

$$Max(|I_{Mn}|) \approx C_m \cdot 2 \cdot V_{DD}/t_{R/F} \approx 0.5 \times 2 \times 1/10 = 0.1$$

where $t_{R/F} \approx 10$ from Figs.1.13.2(a) and (b). This estimate agrees reasonably well with the numerical result of Fig.1.13.4.

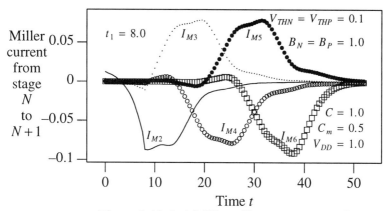

Figure 1.13.4 A Miller-effect current waveform

The current that flows into the Miller-effect gate-field from the driving-voltage signal source at the left end is given by

$$I_{M1} = C_m (d/dt)(V_0 - V_1)$$

This current depends on the rise time of the input voltage V_0, as we saw in Fig.1.13.2(a) and (b). V_0 was a stepfunction from 0 to V_{DD} with a linear-voltage-rise region having duration t_1. I_{M1} versus time for various t_1 is shown in Fig.1.13.5, in a linear time scale and logarithmic current scale. I_{M1} has many features. If the rate of increase of V_0 is high (like the cases of $t_1 = 0.5$ or 1.0), the strong initial current that charges the first C_m in the polarity of node 0 positive and node 1 negative flows.

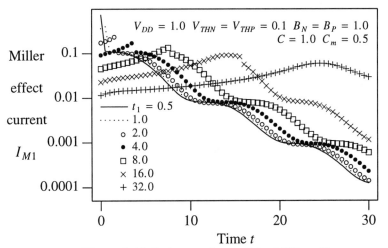

Figure 1.13.5 A waveform of the Miller-effect current at the driving end

This current is a short, impulsive current having duration of the order of t_1. After that, node 1 voltage decreases, and an approximately constant current flows until node 1 is pulled down. If the rate of increase of V_0 is small, the two processes takes place concurrently, as is observed from the cases for $t_1 = 4.0$ to 32.0. The almost constant Miller current continues, again, to the time when pulldown of node 1 is finished. In addition to the conspicuous first-stage pulldown effect, I_{M1} has many less conspicuous features that are caused by the switching process of the second, third, fourth, and so on stage inverters. The Miller current injected by the *remote* stages from the driving end is, however, attenuated by the C-C_m capacitive ladder from the right to the left and does not affect I_{M1} significantly. Approximately exponential dependence of I_{M1} reflects the exponential attenuation effect of the capacitive voltage divider. The attenuation ratio is $C_m/(C_m+C) = 1/3$ per stage. Since the delay time of the inverter is about 5 from Figs.1.13.2(a) and (b), the current decreases approximately to 1/3 for every increment of t by 5. This agrees well with the result of Fig.1.13.5. I_{M1} for $t_1 = 32$ shows almost time-independent Miller current in the time region shown, but this is because the range of the time shown in Fig.1.13.5 is limited to $t < 30$.

In the analysis of the Miller-effect gate-field, it may appear

Classical Gate Field

cumbersome to study a long, yet not too long, inverter chain by solving for the derivatives of the node voltages by the tedious algebra. It might appear that the numerical analysis can be carried out by using digitization of the node equation written in the form

$$C\frac{dV_n}{dt} = \Delta I_n + C_m \frac{d}{dt}(V_{n+1} + V_{n-1} - 2V_n)$$

This scheme works for small C_m. For a large C_m, however, divergence of the node voltage such as that shown in Fig.1.13.6 shows up at large t.

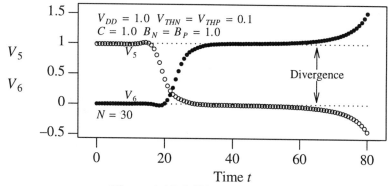

Figure 1.13.6 Divergence of node voltage by improper computational processes

The divergence is a mathematical artifact originating from the simplification used in the numerical analysis, and it does not reflect anything physically meaningful. We note that $V_{n+1} + V_{n-1} - 2V_n$, which is the variable differentiated by t on the right side of the above equation, is proportional to the second partial derivative of the voltage profile, V_n by the spatial coordinate n. Once the second derivative begins to increase, the effect becomes catastrophic, and the computation becomes altogether impossible. Since we wish to cover the entire range of the values of C and C_m, we used a mathematically cumbersome but physically reliable method.

1.14 Feedforward Excitation Transmission

A gate-field may have a feedforward path to increase the propagation velocity of excitation. Of the three circuit components L, C, and R, only C is able to transmit a voltage change of one node to the other node (that is also loaded by capacitance) without delay. The Miller-effect gate-field discussed in Section 1.13 has capacitors connecting the input node and output node of each inverter. Since the input node and output node of an inverter have opposite transition polarities, the connection does not create a constructive feedforward path. A feedforward path can be created by connecting capacitors to the alternative nodes, as is shown in Fig.1.14.1. Feedforward capacitors C_m connect the odd- and the even-numbered nodes separately and in sequence. Since even-numbered nodes have the same transition polarity, and the same is true for odd-numbered nodes, C_m transmits a signal forward constructively. How doe constructive feedforward increases the signal transmission? We need a reference for the node capacitance that determines delay time. If the circuit is observed from node n and if nodes $n-2$ and $n+2$ are both in a quiescent state, then the effective capacitive load of node n, C_{EFF}, is

$$C_{EFF} = C + 2C_m$$

This is the reference-capacitive load of the node to compare the delays. Since the nodes are two inverters apart, the effects of simultaneous switching are less significant than the Miller-effect field of the last section.

We investigate the 8-stage cascaded inverters with the capacitive feedforward coupling, shown in Fig.1.14.1. Again, we consider a long, but not too long, chain. The set of node equations is written as

$$C\frac{dV_1}{dt} = \Delta I_1 + C_m \frac{d}{dt}(V_3 - V_1)$$

$$C\frac{dV_m}{dt} = \Delta I_m + C_m \frac{d}{dt}(V_{m-2} + V_{m+2} - 2V_m) \quad (2 \leq m \leq 6)$$

$$C\frac{dV_m}{dt} = \Delta I_m + C_m \frac{d}{dt}(V_{m-2} - V_m) \quad m = 7, 8$$

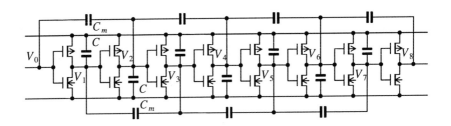

Figure 1.14.1 Feedforward excitation transmission

If we use notation $dV_m/dt = X_m$ as before and transfer the unknowns to the left side of the equations, we get a set of simultaneous equations:

$$(C + C_m)X_1 - C_m X_3 = \Delta I_1$$

$$(C + 2C_m)X_2 - C_m X_4 = \Delta I_2 + C_m X_0$$

$$-C_m X_{m-2} + (C + 2C_m)X_m - C_m X_{m+2} = \Delta I_m \quad (3 \leq m \leq 6)$$

$$-C_m X_{m-2} + (C + C_m)X_m = \Delta I_m \quad (m = 7, 8)$$

If we define
$$Q_A = C^4 + 7C_m C^3 + 15C_m^2 C^2 + 10C_m^3 C + C_m^4$$
$$Q_B = C^3 + 6C_m C^2 + 10C_m^2 C + 4C_m^3$$
$$Q_C = C_m(C^2 + 4C_m C + 3C_m^2)$$
$$Q_D = C_m^2(C + 2C_m) \quad Q_E = C_m^3 \quad Q_F = C_m^4$$

we have
$$X_8 = \frac{Q_B \Delta I_8 + Q_C \Delta I_6 + Q_D \Delta I_4 + Q_E \Delta I_2 + Q_F X_0}{Q_A}$$

and
$$Q_1 = C^4 + 6C_m C^3 + 10C_m^2 C^2 + 4C_m^3 C$$
$$Q_2 = C^3 + 5C_m C^2 + 6C_m^2 C + C_m^3$$
$$Q_3 = C_m(C^2 + 3C_m C + C_m^2)$$
$$Q_4 = C_m^2(C + C_m) \quad Q_5 = C_m^3$$

Then we have
$$X_7 = \frac{Q_2 \Delta I_7 + Q_3 \Delta I_5 + Q_4 \Delta I_3 + Q_5 \Delta I_1}{Q_1}$$

From X_8 and X_7 the other derivatives are computed by using the formula
$$X_6 = \frac{(C+C_m)X_8 - \Delta I_8}{C_m} \quad X_5 = \frac{(C+C_m)X_7 - \Delta I_7}{C_m}$$
$$X_m = \frac{(C+2C_m)X_{m+2} - C_m X_{m+4} - \Delta I_{m+2}}{C_m} \quad (m = 4, 3, 2, 1)$$

Using these formulas the evolution of the node voltages can be determined by a computer. Figure 1.14.2 shows the numerical results, maintaining $C_{EFF} = C + 2C_m = 2$ and varying C and C_m subject to the constraint. $C + 2C_m$ is held constant at 2.0. If C_m is small, as in Fig.1.14.2(a), the node-voltage waveforms are like an ordinary inverter chain. At $C = 1.8$ and $C_m = 0.1$, the cumulative delay of the eight stages is about 36. As C is decreased and C_m is increased, the effective load capacitance is the same, but the cumulative delay decreases. At $C = 1.2$ and $C_m = 0.4$, the cumulative delay decreases to about 31, and at $C = 0.2$ and $C_m = 0.9$ to about 15. The effects of feedforward in increasing the propagation velocity are significant.

As C decreases and C_m increases, the part of the waveform immediately after $t = 0$ changes. The even-numbered nodes make small upward steps immediately following the input transition, and the main pullup of node 2 starts later, on the top of the initial upward step. In node n (even numbered), the pullup of nodes 0, 2, ..., $n-2$ create successively higher steps, and the main pullup transition starts later on the top of the previously accumulated steps. This is most clearly observable from Fig.1.14.2(c). This successive step formation is due to the feedforward effects of the gate-field.

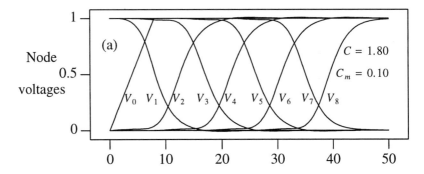

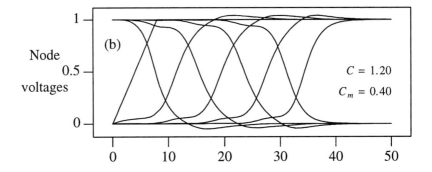

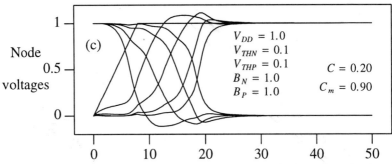

Figure 1.14.2 Feedforward transmission of excitation

As for the odd-numbered nodes, the small downward steps are generated following the pulldown of node 1, since the transition is transmitted from the stage 1 inverter to the inverter stages to the right. The effects of

Classical Gate Field

accumulating of the initial steps are also observable. The feedforward effect is to prepare the nodes to expedite the main transition that arrives later.

The node waveforms have overswing, after the main part of the pullup (even-numbered nodes) or pulldown transition (odd-numbered nodes) is over. After node n transition is over, node $n+2$, $n+4$, ..., node transitions having the same polarity takes place, and they inject current back to node n, thereby creating the overswing. The overswing is more significant at a left-side node than at a right-side node, since there are more nodes to the right for a left-side node, and the effects from all the right-side nodes having the same transition polarity are added together. The fine structures due to the successive transitions are observable in the overswing.

Figure 1.14.3 shows the current J_n injected to node n from nodes $n-2$ and $n+2$ defined by

$$J_n = C_m \frac{d}{dt}(V_{n-2} + V_{n+2} - 2V_n)$$

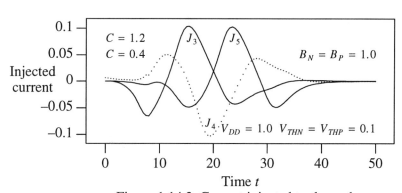

Figure 1.14.3 Current injected to the nodes

The current injected to an odd-numbered node has one peak and two valleys. The peak corresponds to the pulldown of the node itself and is high. The valleys are shallower in depth than the height of the peak. The valley to the left is due to the pulldown of node $n-2$, which takes place sometime before, and the valley to the right is due to the pulldown of node $n+2$ which takes place at the same interval later. The nodes $n-2$, n, and $n+2$ pull down but not at the same time. Therefore, the effects of the pulldown of node $n-2$ do not help the pulldown of node n very effectively. The fine structures observable in the injected current waveform of J_5 at about $t=8$ are due to the pulldown of node 1, and structures observable in J_3 at $t=30$ are due to pulldown of J_7. They are both too far away from the main transition to effect significant speedup. If the pulldowns of the other nodes did take place simultaneously, there would be a quite significant speedup. If the delay time of the inverter is shorter, the capacitive coupling is more effective. This is, however, practically difficult to accomplish, since digital gates work at the speed limit of the processing technology used to build the FETs.

Figure 1.14.4 shows the current injected from the input signal source to the feedforward gate-field, defined by

$$I_0 = C_m \frac{d}{dt}(V_0 - V_1)$$

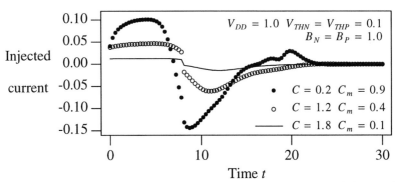

Figure 1.14.4 Current injected from the driving end

I_0 is positive at the beginning, when V_0 is increasing. In this study we assumed

$$V_0(t) = 0 \quad (t < 0) \quad = V_{DD}(t/t_1) \quad (0 \le t \le t_1)$$
$$= V_{DD} \quad (t > t_1)$$

and $t_1 = 8$, the typical rise-fall time of the node. I_0 is positive for $0 \le t \le t_1$, but later as the successive stages go through the upgoing transition, I_0 changes sign. If I_0 is integrated over time, the total charge is zero, since at the beginning of the process, and at the end of the process the capacitor charge is zero. As we see from Fig.1.14.4 for $C = 0.2$ and $C_m = 0.9$, the capacitor current becomes negative, and when this happens, V_0 is still increasing. The input impedance at this time is equivalent to a negative capacitance. A negative capacitance is an effect associated with the capacitive feedback across a noninverting buffer having a voltage gain (Shoji and Rolfe, 1985).

The effect of the negative capacitance can be measured by the duration of time when V_0 is increasing and I_0 is negative. Figure 1.14.5 shows the effects for the three input rise times $t_1 = 8, 16$, and 24. For $t_1 = 8$, the interval is short, $6.9 < t < 8.0$. For $t_1 = 16$, the interval is $11.4 < t < 16$, and for $t_1 = 24$, it is $15.9 < t < 24.0$. The negative capacitance effect is more pronounced for the slower input-voltage transition. From this observation, we conclude that if a significant improvement in the transition delay is required, a feedforward is not adequate.

Capacitance of each node of the gate-field must be reduced either by proper design or by the active capacitance cancellation.

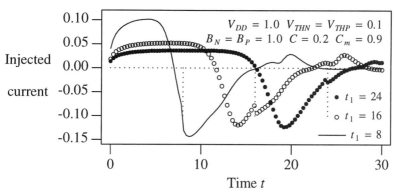

Figure 1.14.5 Current injected from the driving end

1.15 The Gate Field of a Negative-Resistance Diode

A gate-field built using negative resistance diodes is able to support unattenuated propagation of a digital signal. The chain circuit shown in Fig.1.15.1 is built using N-type, or voltage-controlled negative-resistance diodes, like tunnel diodes. The structure of the chain is similar to the gate-fields discussed before, in that the power supply V_{DD} provides energy to each node of the negative-resistance diode field through load resistance R_0. Each node location is loaded by capacitance C, whose charge and energy are the observables of the negative-resistance diode field. Each location is connected to the neighbor locations by coupling resistor R_1. The internode coupling mechanism is different from the FET-based gate-field: the field is able to propagate signals both to the right and to the left, and therefore the field is not unidirectional.

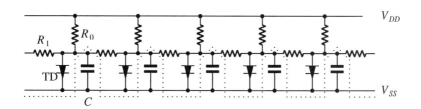

Figure 1.15.1 Tunnel-diode signal-propagation line

The negative-resistance diode field works as follows. Figure 1.15.2 shows the current-voltage characteristic of an N-type negative-resistance diode, like a tunnel diode. A real negative-resistance diode has only *differential* negative resistance and not an absolute negative resistance: only in a part of the I-V characteristic the slope of the curve does dI/dV become negative. Suppose that the gate-field is uniform and that all the nodes of the gate-field of Fig.1.15.1 are at the same voltage. Since no current flows through R_1, each node location is isolated, and the diode and R_0 are effectively connected in series in this state. Let the power-supply voltage be V_{DD}. Then the possible node voltage is determined by the intersection of the I-V characteristic of Fig.1.15.2 and a straight line

$$V = V_{DD} - R_0 I$$

which is conventionally called a *load line*. There are two stable bias points V_L and V_H. V_L is reached if all the capacitors are completely discharged, V_{DD} is held at zero to establish the equilibrium, and then it is increased. V_H is reached if V_{DD} is held at a much higher level than V_H for some time and is reduced to the normal level.

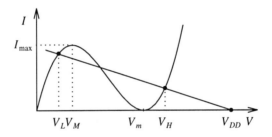

Figure 1.15.2 An I-V characteristic of a tunnel diode

Let the N-type negative-resistance diode characteristic be approximated by a simple cubic polynomial of the form

$$I = AV(V - V_m)^2$$

and simulate the signal propagation through the gate-field. From the formula,

$$V_M = V_m/3 \quad \text{and} \quad I_{max} = (4/27) A V_m^3$$

The parameter values are taken as

$$V_m = 1.0 \quad I_{max} = 1.0 \quad V_{DD} = 5.0 \quad \text{and} \quad N = 30$$

where N is the length of the chain. The circuit parameter values are chosen as $C = 1.0$, $R_0 = 5.0$, and $R_1 = 1.0$. The circuit equation is

$$C \frac{dV_n}{dt} = -I(V_n) + \frac{V_{DD} - V_n}{R_0} + \frac{V_{n+1} + V_{n-1} - 2V_n}{R_1}$$

The static operating point is determined by simulating the biasing procedure described before: by setting $V_{DD} = 0$ at the beginning and then increasing V_{DD} normally, the bias point V_L is reached. As the DC bias is established, the voltage of the left-end node is raised to a high level. Then the signal propagates from the left to right, as is shown in Fig.1.15.3.

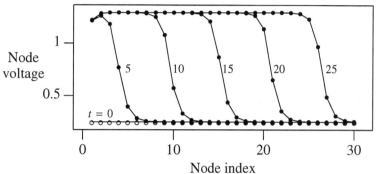

Figure 1.15.3 Voltage profile development in an active RC chain

The voltage profiles from $t = 0$ to $t = 25$ are plotted. The high and the low logic levels are

$$V_H = 1.2917 \quad \text{and} \quad V_L = 0.2504$$

Figure 1.15.4 shows how the velocity of propagation of excitation depends on the circuit parameters. The velocity is primarily determined by the time required to transfer information from location n to $n+1$, and that is time constant $R_1 C$. The velocity is approximately proportional to the inverse of the time constant.

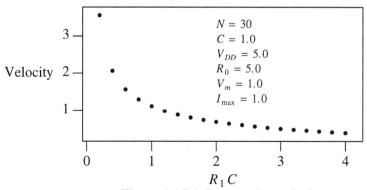

Figure 1.15.4 Propagation velocity versus an interstage coupling-time constant

Chapter 2

Quantum Mechanics of Digital Excitation

2.01 Introduction

The characteristics of the cascaded digital gates in a large-scale chain or mesh structure, where the switching transients develop by traveling signal fronts, are examined in Chapter 1. In this chapter, I proceed to study the characteristics of the transients created in the gate-field, which I call *excitation*. The most fundamental attribute of an excitation is that it is an undividable whole and has its own identification. Particles and waves in three-dimensional physical space are examples of excitation. An excitation carries energy, information, and, often, mass. Thus, digital signals propagating in a logic-gate chain are excitations created in the gate-field. The nature of an excitation reflects the background properties of the field. Since I stressed the similarities between physical space and the gate-field in Chapter 1, here I begin by pointing out the differences between the two. The differences are crucial to an understanding of digital excitation in digital circuits. There are four fundamental differences between physical space and the digital gate-field, which reflect back to the excitations they support:

(1) The physical space is continuous with respect to any of the three spatial coordinates, down to the presently accessible limit, but the digital-gate field is discrete, having the integer node-location index.

(2) A point in physical space exercises a noninverting influence on its neighbor point, but the digital-gate field may exercise either inverting or noninverting influence on its immediate neighbor, depending on the structure of the individual point of the gate-field. This is an important resource for adding variety and flexibility to the gate-field model.

(3) Physical space appears to allow an infinitely large dynamic range of field variables, but the digital-gate field allows only a limited dynamic range of the gate-field variables. The gate-field has a power supply as the energy source that does not allow the voltage variable to exceed the range set by it. Certain field variable values are asymptotic values: An idealized excitation in a finite range may extend over the entire range of

the gate-location index, and as the gate location index tends to plus or minus infinity, the field variable takes the asymptotic values at the infinities.

(4) An excitation in a physical space moves in any direction of physical space, but a digital excitation in a conventional structure moves only in one direction. To make it bidirectional, a certain structure and definition are required (see Section 1.03).

These differences may be considered as the peculiarities of the gate-field, and they are the starting points of our investigation in this chapter.

2.02 Elementary and Composite Excitation

What are the characteristics of an elementary excitation? A gate-field supports two fundamental types of elementary excitation. An elementary excitation is an event that takes place only once in history or in the entire time interval $-\infty < t < +\infty$. This is quite a significant constraint originating from an idealization. If the elementary excitation is observed at a node location, it can be either a LOW to HIGH, or a HIGH to LOW node-voltage transition. The two transition polarities identify the two different types. An elementary excitation generally extends over the entire digital-gate field, and its history depends on many details, such as where it is created, how the gate-field is set up at the beginning, and so on. If the gate-field is infinitely long and uniform (or consisting of the same gates in the same conditions), the elementary excitation asymptotically approaches a stable final voltage profile that reflects the properties of the gate-field. In these respects, there is a significant parallel to common excitations in physical space. The space we live in supports only one kind of electron and positron. They correspond to digital excitation in the steady state. The elementary excitation that has not reached the steady state may be considered as the excited states of the elementary excitation in the field, as they are in quantum mechanics. In the steady state an elementary excitation maintains itself forever, or it cannot decay. A single elementary excitation cannot decay, but two elementary excitations may interact and decay. The decay product is nothing, or a featureless uniform voltage level, the HIGH or the LOW, as I show in the later part of this chapter.

At every location in the gate-field, the Boolean-level switches when the feature of the elementary excitation passes the location. Any digital excitation in the field can be expressed by a superposition of the elementary excitations launched at the periphery of the field or at a launch site within the field, by providing such a peculiar site. If the superposed excitations are observed at a location, the types of elementary excitations that arrive there alternate: a LOW to HIGH transition is always followed by a HIGH to LOW transition and vice versa, since only two levels are recognizable in a binary logic circuit. The switching activities in the

Quantum-Mechanical Gate Field

digital gate-field is understood as a sumtotal of the elementary excitations originated at different places at different times, and they interact each other. A superposition is an operation to build a complex object from simple objects, and it is often difficult to see the consequences of many operations that took place at many locations at many different times. The two simplest examples are shown in Figs.2.02.1 and 2.02.2.

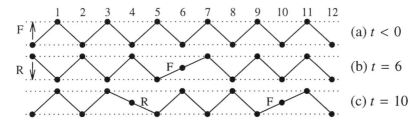

Figure 2.02.1 A voltage profile in a gate-field
of cascaded inverting buffers

Figure 2.02.1 shows an elementary excitation-voltage profile in an inverting buffer field. For $t < 0$, the profile is an alternating HIGH-LOW pattern, where the left-end input node is LOW, the even-numbered index nodes are LOW, and the odd-numbered index nodes are HIGH. The left-end node voltage makes a LOW to HIGH transition at $t = 0$, thereby launching the first elementary excitation. At $t = 6$, the excitation moves six steps rightward and creates a 180-degree phase shift of the HIGH-LOW alternating pattern at F (F stands for *front*), as shown in Fig.2.02.1(b). At $t = 6$, the left-end node voltage makes a HIGH to LOW transition. This transition launches the second elementary excitation, which propagates and creates a phase-shift pattern R (R for *rear*) in the profile at $t = 10$. During that time the front makes ten steps rightward, and the rear end moves by 10 - 6 = 4 steps. Figure 2.02.1(c) shows the simplest composite excitation, a pair of different polarity transitions propagating the constant distance apart between them. The $t = 10$ profile has the same HIGH-LOW pattern as the $t < 0$ profile in the upstream (nodes 0 to 3) and downstream (nodes 11 to 12) regions. The region between R and F (nodes 5 to 9) is the *feature* of the composite excitation in the inverting buffer field, where the HIGH-LOW pattern is locally reversed from the rest.

Figure 2.02.2 shows an elementary excitation in a noninverting buffer field. A noninverting buffer is a cascaded two-stage inverter, that may or may not have the internal-node capacitive loading, as was discussed in Section 1.02. The profile at $t < 0$ is the uniform LOW level as is shown in Fig.2.02.2(a). At $t = 0$, the input of the gate-field, the left end, is pulled up. At $t = 6$, the wavefront propagated six steps to the right, as shown in Fig.2.02.2(b). At that time the input is pulled down. A pair of the downgoing and the upgoing excitations, F and R, respectively, propagate the gate-field. The voltage profile at $t = 10$ is shown in Fig.2.02.2(c).

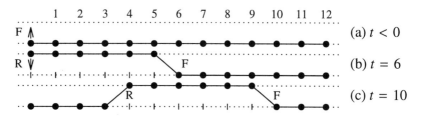

Figure 2.02.2 A voltage profile in a gate-field of cascaded noninverting buffers

The state of a ringoscillator that is discussed in Chapter 3 generally consists of multiple transition edges in the loop that circulate a closed signal path. The excitation in a ringoscillator is a slight modification of the elementary excitation in an infinitely long field. A mathematical technique called the *cyclic boundary condition* modifies the theory of elementary excitation in the infinitely long gate-field to a ringoscillator. Whether a long path or a closed path, any state of a digital circuit having a constant power-supply voltage can be described by the superposition of a set of the elementary transition edges in the respective domain. This is very similar to describing a multiparticle system in terms of so many single particles. In building up the state of a circuit from elementary excitations, there is a fundamental rule. The excitation that passes a point of a digital field in sequence must have alternating transition polarity. From the polarity rule, a pair of upgoing and downgoing excitations traveling together in a gate-field acquires the special status of being a fundamental excitation. In quantum mechanics a similar situation exists: a bound pair of a particle electron and an antiparticle positron make an atom like bound state called a *positronium*. This is the simplest composite excitation in a buffer field, a return-to-zero pulse, shown in Figs.2.02.1(c) and 2.02.2(c).

2.03 Finite and Infinite Energy Associated with Excitation

In a gate-field, the only observable variables are the energies or the voltages of the nodes. Then, a downgoing stepfunction excitation propagating through an infinitely long noninverting gate-field whose Boolean-level profile is shown in Fig.2.02.2(b) gives an unusual impression to the observer: energy is supplied at a constant rate from the driving source, the energy and the excitation move in the same direction, and the energy associated with the excitation increases proportionally to the propagation time. In the limit as $t \to \infty$ the excitation moves away to positive infinity, and the energy diverges. This is against the principle that the observer conventionally subscribes to that the physical world should not demand infinite energy just to propagate the most elementary excitation. This difficulty does not appear if the gate-field has finite length. It is the idealization to an infinite length, which creates the divergence. An upgoing stepfunction in the noninverting gate-field shown by R of Fig.2.02.2(c) gives another unusual impression that the field initially holds infinite energy, and the energy flows out at a constant rate from the driving end. The directions of motion of energy and of the excitation are opposite.

A stepfunction propagating in an infinitely long inverting gate-field has a similar problem, with somewhat different details. The gate-field holds infinite energy at the beginning. Every other node of the gate-field holds energy $(C/2) V_{DD}^2$. Since the number of nodes is infinite, the field holds infinite energy in any of the two steady states. This is the same difficulty that we saw in the noninverting gate-field. The explanation of excitation propagation can be as follows. The propagating wavefront carries energy from one node that holds energy at present to the next node that holds no energy at present, and the energy the entire gate-field holds is the same infinity. As the driving end of the gate-field is driven up, a unit energy, $(C/2) V_{DD}^2$, flows into the gate-field from the signal source. If the driving end is driven down, the same amount of energy flows out of the gate-field: the energy of the gate-field goes back to the signal source. The gate-field exchanges energy to and from the signal source, but the energy is still infinity.

The observations show that diverging energy either the initial energy or the energy associated with the excitation as it propagates is a commonplace phenomenon in an idealized, infinitely long gate-field and its *restricted* observation. Of course, if the background mechanism of the gate-field is included in the consideration, this is nothing surprising: the background circuit indeed consumes infinite energy, either to initialize the nodes or to propagate the excitation. This route of resolution of the divergence is not allowed, however, since we deal with observable variables only. Then the issue becomes a peculiarity, at the least, and a defect, at the worst, of the gate-field model, like the divergence problems of quantum field theory (Aitchison, 1982). There is one way out from this difficulty in a noninverting gate-field: it is to exclude a simple stepfunction from the fundamental excitation of the gate-field. According to this view, the only rationally observable form of excitation is a combination of a downgoing stepfunction followed by an upgoing stepfunction, as shown in Fig.2.02.2(c). The excitation has finite observed energy defined from its voltage profile. None of the difficulties mentioned before exists.

In spite of the rationality, is it a proper procedure to restrict the structure of excitation to this extent? As we see later in Section 2.17, a combination of upgoing and downgoing elementary excitations is not stable, and it decays as it propagates. I seriously doubt that such a temporary object is really elementary. Furthermore, complete exclusion of the excitation in an inverting-gate field, which is more fundamental, is most unconvincing. We had better consider that divergence is real, at the best, and meaningful, at the least. Of the two energy problems, the diverging initial energy of the gate-field is less problematic than the time-dependent and potentially infinite energy associated with a moving excitation. This is because the zero reference of energy is always debatable. If we do conclude and admit that only the energy difference is available to the field to display a phenomenon in it, the issue is resolved in the inverting-gate field. By this interpretation it is possible to include the inverting-gate field in consideration.

We admit that the diverging energy associated with a gate-field is fundamental. The difficulty occurs both in the inverting- and noninverting-gate field in the same way. The solution mentioned before to consider a pair of stepfunctions or RZ pulses as to be the practical but not basic building block and anything simpler than that like a stepfunction, is a fundamental and theoretical construct, is perhaps one rational interpretation, in the following sense. The theory does not have to explain the diverging-energy issues, once the energy difference and not energy itself is involved in the problem. The elementary stepfunctions are like quarks that cannot be observed in an isolated state, in practical information-processing circuits. In the fundamental level, however, its infinite energy resists any influence attempting to destroy it, so it is a permanent existence contrary to the isolated pulse. A logic system never works with a single stepfunction excitation: if a single elementary excitation persists, no processing is carried out. They use composite excitation all the time to process data. This is yet another aspect of compromise in the conventional digital circuit, and we are concerned with this issue because the real digital circuit is never fundamental in its design or its operation. Unless we have a better alternative, we need to stick with this interpretation. We note that in a gate-field it is quite natural to express a RZ pulse as a superposition of the two, time-shifted pair of stepfunctions, each carrying time-dependent energy, and in this case the observable energy is in the form of infinite A minus infinite B equals finite. It is based on the view that the constant energy that the object carries is the difference between the two infinite energies and that such an arithmetic is possible because the invisible background mechanism exists. If I do stretch the imagination, even in physical space there is a background energy source that lends and collect energy. Then isn't it difficult to lend or borrow too much energy for too long a time, and shouldn't that requirement be the basic law of physics?

2.04 An Eigenvalue Problem in the Gate Field

A simple inverting buffer field is made of the identical inverters cascaded over many stages. Figure 2.04.1 shows a section having two stages. The node voltages V_0, V_1, and V_2 are time-dependent waveforms. If the gate-field is infinitely long, and if we consider that there are practically an infinite number of stages to the left side of the section, the node waveforms V_0 and V_2 are the functions of time t satisfying

$$V_2(t) = V_0(t-\tau) \tag{2.04.1}$$

This means that V_2 is a time-shifted replica of V_0. The time shift τ is a cumulative delay of the two inverter stages. If the inverters are pullup-pulldown symmetrical, $V_1(t)$ becomes an inverted and then time-shifted replica of $V_0(t)$ defined by $V_1(t) = V_{DD} - V_0(t-\tau')$, as was discussed in Section 1.06. We consider here the inverters to be identical but not pullup-pulldown symmetric.

Mathematically the requirement can be stated as follows. The gate-field is described by a set of infinite number of differential equations of the form

$$C\frac{dV_n}{dt} = \Delta I(V_{n-1}, V_n) \tag{2.04.2}$$

$$C\frac{dV_{n+1}}{dt} = \Delta I(V_n, V_{n+1})$$

where $\Delta(V_{n-1}, V_n)$ is the current that the pair of the FETs of the n-th stage inverter jointly inject to the node n capacitor. According to Eq.(2.04.1) in the limit as $n \to \infty$, we have the solution satisfying

$$V_{n+1}(t) = V_{n-1}(t-\tau)$$

for a certain value τ. This problem is an eigenvalue problem of the infinite set of differential equations: all the $V_{n\pm1}(t)$ are the eigenfunctions, and τ is the eigenvalue.

The mathematical difficulties in solving this problem originate from the complexity of the function $\Delta(V_{n-1}, V_n)$, which is traced back to the complexity of the FET I-V characteristics. Generally, a closed-form solution of the problem is not available. There is one special case that can be solved in a closed form. This is the case where the current-voltage characteristics of the FETs are given by the collapsable current-generator model Shoji, 1992). In this section a generalization of the analysis in (Shoji, 1992) to a pullup-pulldown asymmetrical inverter field is presented. By the model the I-V characteristic of an NFET is given by

$$I_N = 0 \ (V_G < V_{THN}) \quad I_N = I_{max} \ (V_G > V_{THN} \text{ and } V_D > 0)$$
$$V_D = 0 \ (V_G > V_{THN} \text{ and } I_N < I_{max})$$

and where $I_{max} = G_{mN}(V_G - V_{THN})$ is the saturation current of the NFET. The PFET characteristic can be derived by substituting $I_N \to I_P$, $V_D \to V_{DD} - V_D$, $V_G \to V_{DD} - V_G$, and $V_{THN} \to V_{THP}$. We solve the eigenvalue problem in this section, using the simplified I-V characteristic. For further simplification we assume that $V_{THN} = V_{THP} = 0$. We use Figs.2.04.1 and 2.04.2 as a guide to the following analysis.

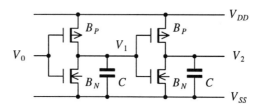

Figure 2.04.1 An engenvalue problem of an inverter field

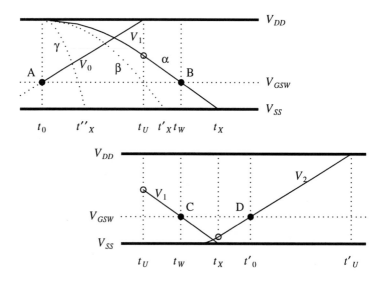

Figure 2.04.2 A guide to solving the eigenvalue problem

The switching threshold voltage of the inverter, V_{GSW}, is given by

$$V_{GSW} = G_{mP} V_{DD}/(G_{mN} + G_{mP}) \qquad (2.04.3)$$

With reference to Fig.2.04.2, suppose that at $t = t_0$, $V_0(t_0) = V_{GSW}$. The t-dependence of $V_0(t)$ for $t < t_0$ is not known yet, but it does not matter. The unknown time domain is indicated by the dotted curve in

Fig.2.04.2. We assume that $V_0(t)$ is given by

$$V_0(t) = V_{GSW} + (G_{mP} V_{DD}/C)(t - t_0) \quad (2.04.4)$$

in the region $t > t_0$. We note that this assumption *must be* checked if it is really valid after a few more steps of the analysis are carried out. $V_0(t)$ reaches V_{DD} at $t = t_U$, given by

$$t_U = t_0 + \frac{C G_{mN}}{G_{mP}(G_{mN} + G_{mP})} \quad (2.04.5)$$

Node voltage V_1 satisfies the circuit equation

$$C \frac{dV_1}{dt} = G_{mP}(V_{DD} - V_0) - G_{mN} V_0$$

$$= G_{mP} V_{DD} - (G_{mN} + G_{mP}) V_0$$

$$= -(G_{mN} + G_{mP})(V_0 - V_{GSW}) \quad (t > t_0)$$

By substituting Eq.(2.04.4) and using the initial condition $V_1(t_0) = V_{DD}$, we obtain

$$V_1(t) = V_{DD} - \frac{G_{mP}(G_{mN} + G_{mP}) V_{DD}}{2C^2}(t - t_0)^2 \quad (2.04.6)$$

Equation (2.04.6) is valid if $t < t_U$. If $V_1(t_U) > 0$, we have for $t > t_U$ and $V_1(t) > 0$,

$$V_1(t) = V_1(t_U) - (G_{mN} V_{DD}/C)(t - t_U) \quad (2.04.7)$$

$$(t > T_U)$$

because $V_0(t) = V_{DD}$ the PFET is turned off and because the NFET pulldown current is constant at $G_{mN} V_{DD}$. There are two cases. With reference to Fig.2.04.2,
(1) Eq.(2.04.6) is valid until $V_1(t)$ reaches 0 at $t = t''_X$, and after that, $V_1(t) = 0$.
(2) Eq.(2.04.6) is valid only in the range $t_0 < t < t_U$, and for $t > t_U$ Eq.(2.04.7) is valid, until $V_1(t)$ given by Eq.(2.04.7) becomes zero at t_X. After that $V_1(t) = 0$.

Equation (2.04.6) becomes zero at

$$t''_X = t_0 + \frac{\sqrt{2} C}{\sqrt{G_{mP}(G_{mN} + G_{mP})}}$$

For $t''_X < t_U$ to be satisfied, we have $G_{mN}^2 - 2 G_{mP} G_{mN} - 2 G_{mP}^2 > 0$, or $G_{mN} > (1 + \sqrt{3}) G_{mP} = 2.732 G_{mP}$. We then test a somewhat stronger condition than this: if $V_1(t)$ of Eq.(2.04.6) equals or is greater than V_{GSW} at $t = t_U$. This condition is satisfied if

$$G_{mN} \leq 2 G_{mP}$$

and if $G_{mN} = 2G_{mP}$, we have $V_{GSW} = (1/3) V_{DD}$. By observing the two conditions, let us consider the case where $G_{mP} \leq G_{mN} \leq 2G_{mP}$, or $(1/2) V_{DD} \geq V_{GSW} \geq (1/3) V_{DD}$. This is a realistic case of a moderately asymmetrical inverter that is still close to the pullup-pulldown symmetry. The other case is straightforward, but the algebra is too complex, and it is not worth presenting the results here.

At $t = t_U$ we have

$$V_1(t_U) = \frac{2G_{mP}^2 + 2G_{mN}G_{mP} - G_{mN}^2}{2G_{mP}(G_{mN} + G_{mP})} V_{DD}$$

and we note $V_1(t_U) > V_{GSW}$ if $G_{mN} < 2G_{mP}$, as we have shown. In the range $t > t_U$, Eq.(2.04.7) is valid. $V_1(t)$ of Eq.(2.04.7) becomes V_{GSW} at $t = t_W$, which is given by

$$t_W - t_U = \frac{C(2G_{mP} - G_{mN})}{2G_{mP}(G_{mN} + G_{mP})} \tag{2.04.8}$$

$V_1(t)$ of Eq.(2.04.7) becomes zero at $t = t_X$, which is given by

$$t_X - t_U = \frac{C(2G_{mP}^2 + 2G_{mN}G_{mP} - G_{mN}^2)}{2G_{mN}G_{mP}(G_{mN} + G_{mP})} \tag{2.04.9}$$

$$t_X - t_W = \frac{CG_{mP}}{G_{mN}(G_{mN} + G_{mP})} \tag{2.04.10}$$

This is, of course, a rational result since this is the ratio of V_{GSW} and the rate of the node 1 voltage decrease [the rate of decrease of V_1 is $G_{mN}V_{DD}/C$ and $V_{GSW} = G_{mP}V_{DD}/(G_{mN} + G_{mP})$ as given by Eq.(2.04.3)]. In the time domain, $t_W < t < t_X$, $V_1(t)$ can be written as

$$V_1(t) = V_{GSW} - (G_{mN}V_{DD}/C)(t - t_W) \tag{2.04.11}$$

The node 2 pullup begins at $t = t_W$. The pullup is determined by the equation

$$C\frac{dV_2}{dt} = G_{mP}V_{DD} - (G_{mN} + G_{mP})V_1$$

$$= (G_{mN} + G_{mP})(V_{GSW} - V_1)$$

and Eq.(2.04.11) as

$$V_2(t) = \frac{G_{mN}(G_{mN} + G_{mP})V_{DD}}{2C^2}(t - t_W)^2 \tag{2.04.12}$$

$$(t > t_W)$$

At $t = t_X$, we have

$$V_2(t_X) = \frac{G_{mP}^2 V_{DD}}{2G_{mN}(G_{mN} + G_{mP})}$$

Quantum-Mechanical Gate Field

We check if $V_2(t_X) < V_{GSW}$. We have

$$\frac{G_{mP}^2 V_{DD}}{2G_{mN}(G_{mN}+G_{mP})} < \frac{G_{mP} V_{DD}}{G_{mN}+G_{mP}} \quad \text{if} \quad (2.04.13)$$

$$\frac{G_{mP}}{2G_{mN}} < 1$$

or $G_{mP} < 2G_{mN}$. Since we assumed $G_{mN} \leq 2G_{mP}$, we have the valid range of the present analysis as follows:

$$(1/2)G_{mP} \leq G_{mN} \leq 2G_{mP} \quad (2.04.14)$$

Then for $t > t_X$,

$$V_2(t) = V_2(t_X) + (G_{mP} V_{DD}/C)(t - t_X) \quad (2.04.15)$$

Since the NFET is turned off, $V_2(t)$ reaches V_{GSW} at t'_0, which is given by

$$t'_0 = t_X + \frac{C(2G_{mN} - G_{mP})}{2G_{mN}(G_{mN}+G_{mP})} \quad (2.04.16)$$

From Eqs.(2.04.14),(2.04.9), and (2.04.5) we have

$$t'_0 - t_0 = \frac{C(G_{mN}^2 + 4G_{mP}G_{mN} + G_{mP}^2)}{2G_{mN}G_{mP}(G_{mN}+G_{mP})} = \tau \quad (2.04.17)$$

This is the eigenvalue, which equals the cumulative delays of the two stages of the generally asymmetric but identical cascaded inverters. From Eq.(2.04.17) we observe that the eigenvalue is symmetrical with the exchange of G_{mN} and G_{mP}, and this is, of course, rational.

In a special case of a pullup-pulldown symmetrical-inverter field, we have $G_{mN} = G_{mP} = G_m$. Then we have the following simplified results:

$$\tau = (3/2)(C/G_m) \quad V_{GSW} = V_{DD}/2$$
$$t_U = t_0 + (2/4)(C/G_m)$$
$$t_W = t_U + (1/4)(C/G_m) = t_0 + (3/4)(C/G_m)$$
$$t_X = t_W + (1/2)(C/G_m) = t_0 + (5/4)(C/G_m)$$

and

$$V_0(t) = V_{GSW} + (G_m V_{DD}/C)(t - t_0) \quad (t_0 \leq t \leq t_U)$$
$$V_1(t) = V_{DD} - (1/2)(G_m/C)^2 (t - t_0)^2 \quad (t_0 \leq t \leq t_U)$$
$$V_1(t) = V_1(t_U) - (G_m V_{DD}/C)(t - t_U) \quad (t_U \leq t \leq t_X)$$

where $V_1(t_U) = (3/4)V_{DD}$, and

$$V_2(t) = (1/2)(G_m/C)^2 (t - t_W)^2 \quad (t_W \leq t \leq t_X)$$

$$V_2(t) = V_2(t_X) + (G_m V_{DD}/C)(t - t_X) \quad (t_X \leq t \leq t'_U)$$

where $V_2(t_X) = (1/4)V_{DD}$. The eigenfunction of this special case is plotted in Fig.2.04.3. This result was derived in my last book (Shoji, 1992). The analysis of this section is a generalization of the simplified theory.

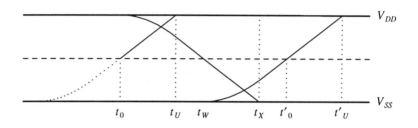

Figure 2.04.3 The eigenfunction in the case of $G_{mN} = G_{mP} = G_m$

2.05 The Eigensolution of a Gate-Field Waveform

In the last section we observed that if the input to a long inverter field built by cascading identical inverters is driven, a wavefront is launched into the field, and after many stages of propagation, the node-voltage waveforms approach the eigenfunction solution. The analysis of the last section is limited to a special FET model of the collapsable current generator (Shoji, 1992). If the FET characteristic is more realistic, such as that given by the gradual-channel, low-field model, we must solve the problem numerically. The simplest way is to simulate a long cascaded inverter chain by a computer. If we do that, we observe the following. By a casual observation of the node waveforms, the convergence limit is practically reached quite soon, after four to six stages of propagation along the gate-field. There are a few interesting details in the final waveform. Here we consider the location index $n \to \infty$ limit to investigate the mathematical details. In the limit, the eigenfunction node waveform looks as shown in Fig.2.05.1 (even-numbered node). We consider the asymptotic expression of the eigenfunction waveform in the limit both as $t \to \infty$ and $t \to -\infty$.

We assume that the FETs are modeled by using the gradual-channel, low-field model. In the limit as $t \to +\infty$, the input voltage to the inverter has already reached the HIGH logic level V_{DD}, and the node equation is well approximated by

$$C \frac{dV_1}{dt} \approx B_P(V_{DD} - V_{THP})(V_{DD} - V_1)$$

The equation has the asymptotic solution

$$V_1(t) \to V_{DD} - A\exp(-t/t_P) \quad \text{where}$$
$$t_P = C/[B_P(V_{DD} - V_{THP})]$$

The approach to the final steady-state level at V_{DD} is an exponential decay process, with the time constant t_P.

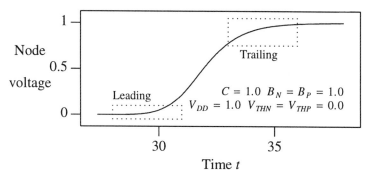

Figure 2.05.1 Leading and trailing edges of the eigenfunction waveform

The limit as $t \to -\infty$ depends on the value of the FET conduction-threshold voltages, V_{THP} and V_{THN}. If $V_{THP} > 0$, the decreasing input voltage $V_0(t)$ depends on time linearly as

$$V_0(t) = V_{DD} - V_{THP} - \alpha t$$

and therefore $V_0(0) = V_{DD} - V_{THP}$, and the relationship holds in the vicinity of $t = 0$. Then the node 1 equation is approximated by

$$C\frac{dV_1}{dt} = (B_P/2)(V_{DD} - V_{THP} - V_0)^2 - B_N(V_0 - V_{THN})V_1$$

If we assume $V_1(t)$ is small and $V_1(t) = \beta t^m$, the left side is $C\beta m t^{m-1}$, and at least one term on the right side of the equal sign must be of the order of t^{m-1}. Then we must have $m = 3$. The second term on the right side becomes negligible by this choice of m. We have an approximate solution

$$V_1(t) = (B_P \alpha^2/6C)t^3 + o(t^4) \quad (t \geq 0) \qquad = 0 \quad (t < 0)$$

as we derived before, in Section 1.05. For a pulldown transition for the initial input voltage $V_0(t) = V_{THN} + \alpha t$, the initial output voltage is given by

$$V_{DD} - V_1(t) = (B_N \alpha^2/6C)t^3 + o(t^4) \quad (t \geq 0) \qquad = 0 \quad (t < 0)$$

and in either case the right side of the equation is precisely zero for $t < 0$.

If $V_{THP} \neq 0$ and $V_{THN} \neq 0$, the initial dependence is $\approx t^3$. If $V_{THP} = V_{THN} = 0$, then we have quite a different mathematical problem. Let the node voltages be V_0, V_1, and V_2 in the order of propagation of the excitation. The equations satisfied by V_1 and V_2 are

$$C\frac{dV_1}{dt} = -(B_N/2)V_0^2$$
$$+B_P[V_{DD}-V_0-(1/2)(V_{DD}-V_1)](V_{DD}-V_1)$$
$$C\frac{dV_2}{dt} = (B_P/2)(V_{DD}-V_1)^2 - B_N[V_1-(1/2)V_2]V_2$$

V_0 increases, V_1 decreases, and V_2 increases. If we set $\phi_1 = V_{DD} - V_1$, and if the small terms in the limit as $t \to -\infty$ are neglected, then we get

$$C\frac{d\phi_1}{dt} = (B_N/2)V_0^2 - B_P V_{DD}\phi_1$$
$$C\frac{dV_2}{dt} = (B_P/2)\phi_1^2 - B_N V_{DD} V_2$$

and we seek for a solution satisfying

$$V_2(t) = V_0(t-\tau)$$

The set of equations cannot be solved in a closed form. To gain insight into the problem, we first assume that the time derivatives are negligible. Then we have a pair of static solutions

$$\phi_1 = [B_N/(2B_P V_{DD})]V_0^2 \quad V_2 = [B_P/(2B_N V_{DD})]\phi_1^2$$

We note here that both ϕ_1 and V_2 increase with increasing time. Then the time derivatives are positive. Then the *static* estimate of ϕ_1 and V_2 are both overestimates. We have

$$V_2 = [B_N/(8B_P V_{DD}^3)]V_0^4 = fV_0^4$$

Let us seek for a function that satisfies this functional relation. Let the function be $\Phi(t)$. It then satisfies

$$\Phi(t-\tau) = f\Phi(t)^4 \quad \text{where} \quad f = B_N/(8B_P V_{DD}^3)$$

We introduce $\Psi(t) = \log \Phi(t)$. This function satisfies

$$3\Psi(t) + \log(f) = \Psi(t-\tau) - \Psi(t) \approx -\tau\frac{d\Psi(t)}{dt}$$

This differential equation can be solved as

$$\Psi(t) = A\exp[-3(t/\tau)] - \frac{\log(f)}{3}$$

and then $\Phi(t)$ is solved as

$$\Phi(t) = f^{-1/3}\exp[A\exp[-3(t/\tau)]]$$

If $\Phi(0) = 1$ for convenience,

$$\Phi(t) = f^{-1/3} \exp[\frac{\log(f)}{3} \exp[-3(t/\tau)]]$$

where $0 < f < 1$, and therefore $\log(f) < 0$. This is an unusually strongly t-dependent function that is not often seen in mathematical physics. The strong dependence is consistent with the observation that the eigenfunction waveform is reached only in several steps of propagation in the gate-field: there are infinitely many stages in the downstream, but only a few stages are sufficient at the upstream of the excitation to represent the eigenfunction with high accuracy.

Figure 2.05.2 shows a numerical result. In this numerical analysis, the time increment of numerical integration was varied to check whether the result is independent on the choice, since the result covers over a wide order of magnitudes. If the simple mathematical form at the leading edge given above is correct, we must have

$$V_2(t-\tau) = [B_N/(8 B_P V_{DD}^3)] V_2(t)^4$$

In Fig.2.05.2, $B_N/(8 B_P V_{DD}^3) = 1/8$. From a numerical simulation, we get $\tau = 3.7$. At $t = 27.1$, node voltage equals 10^{-6}. This means that at $t = 27.1 - 3.7 = 23.4$, V_2 must be $10^{-24}/8 = 1.25 \times 10^{-25}$.

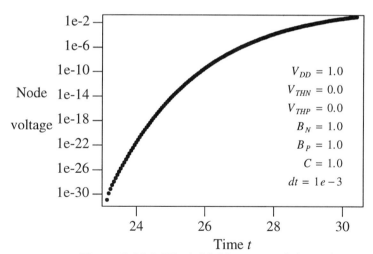

Figure 2.05.2 The initial increase of the node voltage in a gate-field

The numerical analysis gives $V_2 \approx 10^{-28}$ at the value of t. The estimated value is larger than the numerical value. The difference is due to the neglect of $d\phi_1/dt$ and dV_2/dt when we used the static relationship between V_0 and V_2 instead of the hard to solve correct relationship, and thinking of the range of 20 orders of magnitude, the agreement is not unreasonable. The mechanism of creating the leading edge profile of the wavefront is not trivial, but it was explained completely by the theory of this section.

2.06 Gate-Field Variable Measurements

The electrical phenomena in a gate-field are recognized when the field variables, the node voltages, are measured conforming to the laws of the circuit theory. Measurement of a node voltage is not a trivial issue because (1) accessing a node is not trivial, and (2) a measurement process affects the measured node. In my last book, I showed that a voltage measurement is equivalent to a capacitive loading to the accessed node (Shoji, 1996). Accordingly, the voltage of a circuit node that has no capacitance is unmeasurable. The gate-fields of Chapter 1 were constructed to conform the voltage-measurability requirement. If the gate-field is considered to be a macroscopic object, the state of the circuit is characterized by the node voltages that can be measured as precisely as we wish, and the measurement never affects the node. If the gate is considered as a microscopic object, the node voltage can be measured only to one-bit accuracy, the HIGH or the LOW, and therefore the gate-field becomes a Boolean-level field. Analog and digital variable measurements are fundamentally different, since the former is a precise quantitative, and the latter is an imprecise qualitative information acquisition. This point has never been clearly recognized, which has created much confusion. The difference leads to the conclusion that the present digital circuit model is a product of compromise. Analog measurement is classical, and digital measurement is essentially quantum-mechanical. This difference is not well recognized in digital circuit tests since conventional digital circuits are quite compromised objects, very different from mistaken analog-circuit models. The quantum-mechanical measurement of the Boolean-level has many issues that must be clarified.

The fundamental difference between an analog voltage measurement and a Boolean-level determination is described as follows. Analog voltage measurement connects a voltmeter (the simplest model is a quadrant electrometer) to a pair of nodes and reads the angle of deflection of the rotating electrode. The measurement is carried out by an intelligent experimenter (often by a human), and what he does is always considered to be appropriate. The experimenter is expected to exercise caution not to break the laws of physics and to ensure high accuracy of the measured values. He is assumed to have all the capabilities and resources to do so. The measurement is accurate down to the voltage level determined by the laws of physics, especially thermodynamics. In a Boolean-level determination, the intelligent experimenter does not exist. In a logic system, a simple gate circuit carries out the measurement at each step of the digital signal propagation. Accordingly, the measurement is quite unsophisticated. In a digital circuit, every gate is an observer. If digital gates A, B, C, ..., are cascaded in order, Gate B is the observer of gate A output, and gate C is the observer of gate B output, and so on. The subject that carries out the measurement is most often a simple and unsophisticated digital gate or its combination, and at best it is a somewhat more sophisticated piece of digital test equipment, like the automatic IC test machine. The measurement carried out by each logic gate is a low-quality Boolean-level determination. In spite of the observer's simplicity, it is expected to extract a qualitative conclusion, the HIGH or the LOW Boolean-level, from the quantitative voltage waveform. This is quite a demanding task for a simple digital gate circuit. This is the a source of many fundamental problems.

If logic gates A, B, C, ..., are cascaded, logically gate B should process the signal from A after A has completed its Boolean-level determination, and so on, as the signal flows downstream. In practice, before A completes processing, gate B begins processing, and even gate C may get an early warning signal and may begin processing as well. This is an inevitable consequence of using a very simple and unsophisticated circuit for the Boolean-level determination. Such an overlap of the processing interval expedites generation of the final logic answer some of the time, but it has significant problems, both practically and theoretically. Theoretically, the uncertainty of measurement accumulates in the time domain, such as delay-time uncertainty. Practically, if the node voltage change is not monotonous (including a do-undo process), the time required to process the information becomes quite uncertain. Theoretically, it often becomes impossible to determine whether a digital signal moved from A to B or from B to A (Section 1.11). One immediate consequence at this point is a recognition that the delay time of the gate becomes an ambiguous parameter. Generally, an intelligent interpretation based on the observation over the entire logic-gate chain is required even for determination of the direction of propagation (Shoji, 1992). That is the capability of intelligent analog measurement but not of digital. An analog circuit carries out a stereotyped operation, but its outside observer must be intelligent. A digital circuit carries out an imprecise operation, but it has intelligent internal observer.

Boolean-level determination by a logic gate is carried out by the nonlinearity or, more precisely, by the saturability of the gate circuit. The saturation characteristic of an individual gate is not very good, since a CMOS gate as an amplifier has only modest gain. There is a better way to determine the Boolean-level, by effectively increasing the gain of the gates to infinity to improve resolution. This is to use a latch. A latch brings the voltage level into the circuit and then applies positive feedback. In the circuit of Fig.2.06.1(a) the switch is open until the mesurement is initiated. When the switch is closed, positive feedback is applied to the noninverting binary buffer (b) or the the ternary identity buffer (c), and it settles the output voltage to the binary or the ternary quantized level. In this measurement the time when the measurement is initiated is set by the time when the switch closes. This measurement brings one more peculiarity of Boolean-level determination: a continuous measurement is impossible.

In a conventional logic-circuit, Boolean-level determination by the latch is carried out only when the logic-circuit output-voltage levels are transferred to the latches. The latches settle, on a clock transition, at the final logic level, which gives the Boolean-level of the processed data. This is true in a multilevel logic as well. A latch consists of a buffer in a positive feedback loop as shown in Fig.2.06.1(a). The buffer can be a binary buffer [Fig.2.06.1(b)], a ternary buffer [Fig.2.06.1(c)], or any n-ary multivalue logic buffer (Section 3.03): the triangular symbol of Fig.2.06.1(a) can be replaced by either Fig.2.06.1(b) or Fig.2.06.1(c), and Fig.2.06.1(a) becomes a binary or a ternary latch circuit, respectively. In a steady state of a latch at the end of measurement, the output is connected via a conducting FET of the latch circuit, to one of the power-supply buses, from which the unlimited energy required to make the measured Boolean-level useful for further signal processing is available. The measured result has a backup of energy resource to influence the rest of the circuit.

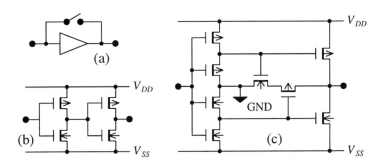

Figure 2.06.1 A digital value-determination circuit

The reason that a latch provides a better performance in quantization is that the signal circulating a closed path of the positive feedback accumulates any high gain to help the circuit reach saturation. The latch settles at a Boolean-level starting from any state, if the settling time is not the issue. This is the another piece of evidence that the measurement time is involved essentially in a Boolean-level determination. A conventional gate may present an intermediate state that is neither the HIGH or the LOW Boolean-level. This is because gain of the gate is used only once, and that can be high but is limited. A feedback creates something qualitatively different, but even this method suffers from uncertainty, as it is discussed in Section 2.10.

2.07 Latch Circuit for Boolean-Level Determination

Let us consider the physical meanings of the process of the Boolean-level determination discussed in Section 2.06. In this exercise, I wish to highlight the crucial differences between the macroscopic (classical) and the microscopic (quantum-mechanical) measurements. The differences between the two may not lead us to an immediate practical application, but it is quite important to understand the fundamental difference to penetrate deeply into the working mechanisms of a digital circuit. Much confusion arises by neglecting the crucial difference because a practical digital circuit operates in a compromised mode, between the classical and the quantum (or idealized) extrema. To see the issues clearly we need to step down to the idealized model of a digital circuit. The idealized model is itself consistent, but is quite different from the conventional digital circuit. That is reached in the limit as the ultimate scaledown of the technology.

The Boolean HIGH and LOW voltage levels of a digital circuit are the quantized voltage levels. To study a digital circuit in its *ideal* model, the difference between the digital HIGH voltage V_{HIGH} and the low voltage V_{LOW} given by

$$\Delta V_{HL} = V_{HIGH} - V_{LOW} \qquad (2.07.1)$$

the minimum *distinguishable* voltage by using the observational method of the Boolean-level. Practically, this limit is reached when the noise existing in the digital system becomes comparable to ΔV_{HL} by the reduction of the power-supply voltage required by the technology scaledown. The FETs are small are working at the low-power-supply voltages, and the circuits are small and sensitive to the perturbation. The effect we discuss here is not a genuine quantum-mechanical effect originating from the small device size. It is strictly the various vulnerabilities of the microscopic circuit. This limit has not yet been reached, but the crucial point is that the idealized model of a digital circuit is established only in this limit, as the opposite extrema of a macroscopic analog circuit. Equation (2.07.1) means that it is inconsistent to introduce any voltage level between the two and to give it valid and consistent physical meanings. An example of such a voltage we know already is the inverter-switching threshold voltage, V_{GSW}. This voltage is between the two logic levels, and it is valid only if the digital circuit is considered as a classical, macroscopic object, whose properties are determined by the *classical* analog measurement. If a digital circuit is considered as an idealized quantum-mechanical object, this freedom must be abandoned. From this restriction, a number of consequences follow. The first is a question what is the significance of the measured value, which is the Boolean-level? and the second is the physical meaning of the gate-delay time, which is closely associated with the logic-threshold voltage. In the following discussions, we consider an idealized digital circuit. The circuit-theoretical problems of measurement in the idealized limit can be stated simply: a measurement perturbs the circuit quite significantly, and after the measurement the circuit may or may not be in the same state as before the measurement.

Suppose that the Boolean-level of node X of a gate-field shown in Fig.2.07.1(a) or (b) is to be determined. The simplest circuit that is able to quantize the voltage level is a latch circuit. The test equipment is a binary latch connected to it, as shown in Fig.2.07.1(a). The circuit enclosed in a dotted rectangle is the latch, which works as the test equipment. In Fig.2.07.1(a), switch SW is closed at the moment when the measurement is initiated. If node X voltage is higher than the noninverting buffer D switching-threshold voltage V_{GSW}, the latch settles at the HIGH logic level, and vice versa. As the latch settles, it enforces its output level to the tested gate-field: the state of node X is now *forced* to the HIGH logic level. The measured state and the state of the measured circuit after the test are the same. This may appear strange, but this is the most fundamental Boolean-level determination of the idealized digital circuit. Yet the arrangement appears from the conventional circuit-measurement practice somewhat strange. Why? The tested circuit is a microscopic object that is unable to deliver a lot of energy to the test equipment. The test equipment borrows a small amount of energy from the node capacitance of the circuit and then returns the borrowed energy to compensate for the damage inflicted before.

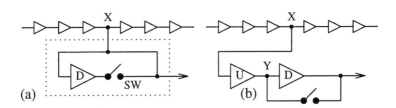

Figure 2.07.1 Boolean-level determination

This is consistent with the assumption that any node that is measurable has capacitive loading. Then this is the minimum perturbation test arrangement of a microscopic object. Borrowing energy and returning, however, it occurs in succession, and not simultaneously. Since the test equipment is able to drive a significant load, the outside observer is able to draw his share of energy from the output and is able to record the test result on some permanent storage. Since the test equipment forces the tested circuit to the state determined by the test, there is no question about the state of the node after the test: the tested circuit is brought into the known state. We note here that the test result was *created* by the test circuit. The difficulty of the test equipment unpredictably influencing the tested node was *not* eliminated but was minimized, and that is the best we can do. A test is a procedure to determine the state of a node, as well as to bring the tested object to the known state (from the known state the tested object goes through evolutionary developments).

A question is that a unity gain buffer could be inserted as shown in Fig.2.07.1(b) (buffer U) to isolate the test equipment. There are several difficulties to accepting this arrangement as fundamental.

(1) Most fundamentally, a unity gain buffer must reproduce the voltage level between the HIGH and the LOW levels faithfully. This is a violation of the assumption we made on an idealized digital circuit. The test equipment is built using the same FETs and the circuit technology as the tested circuit. If such a buffer existed, its input and output has definite zero voltage difference, and this is a violation of the assumption. Logically, the output of the idealized buffer should be in an unknown voltage between the two logic levels.

(2) From (1), a unity gain buffer can be considered as a part of the tested circuit and not of the test equipment. Determination of Boolean-level by the observer requires the Boolean-level of node X and not node Y of Fig.2.07.1(b).

(3) A unity gain buffer is a precision analog circuit like an operational amplifier in a negative-feedback loop. The amplifier has high gain and accordingly narrow bandwidth. Since the buffer U must be built using the same technology to be compatible in the very large scale integrated (VLSI) circuit environment, it cannot be small, fast, and precise simultaneously. Furthermore, the energy drawn by U cannot be returned to node X, and therefore the arrangement affects the state of node X more significantly than the test equipment of Fig.2.07.1(a). The crucial point of the test arrangement of Fig.2.07.1(a) is that the energy drawn for measurement should be returned as quickly as possible but consistently with the response time of the circuit and the test equipment.

(4) Delay through the unity gain buffer does not allow an instantaneous measurement initiated by switch closure. Since a feedback brings the values of the past to the present, its use should be limited to the essential operation of the Boolean-level determination.

(5) The arrangement of Fig.2.07.1(b) is different too much from the *internal* measurements within the logic circuit. Internal measurements are carried out by each gate, as it accepts the driving signal. The results of the internal measurement are temporarily stored as the energy of the output capacitance of the gate. The logic gates between the latches operate in this mode. The gate need not force to the source because the internal observer, the gate, lacks the intelligence to use the data to make a high-level decision. Its task is only to send the data to the downstream gate after processing. The internal measurements are carried out using the node capacitances as the dynamic storage, and as such its results are only conditionally available. Yet it is required that the real test method is closely similar to the internal measurement.

(6) In a conventional synchronous logic circuit, a static latch may be considered as executing an observation in essentially the same scheme as Fig.2.07.1(a), and its results are available without restriction to the next clock-period signal processing.

In the following studies we use the test equipment directly connected to the tested circuit as shown in Fig.2.07.1(a). The test equipment must be represented by an equivalent circuit that is integrated into the tested circuit. The simplest equivalent-circuit model of the test equipment is shown in Fig.2.07.2. In this equivalent circuit, the differential amplifier creates the difference between the input and the reference voltages, and that drives the output FETs, which are modeled by the collapsable current generator. If $V_{IN} < V_{REF}$, the test equipment draws current I_N to ground. If $V_{IN} > V_{REF}$, it injects current I_P. The current generators collapse if the node voltage arrives either at V_{DD} or V_{SS}. The driver stage of the test equipment charges and discharges capacitance C, which is the node capacitance of the tested circuit. The switch at the output closes at the time when the test is initiated.

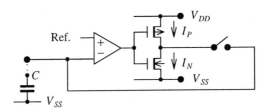

Figure 2.07.2 A model of test apparatus

Setting up a reference voltage between V_{HIGH} and V_{LOW} may appear to be a violation of the assumption. This V_{REF} has a peculiar characteristic that will be discussed in Section 2.09. All the components of Fig.2.07.2 work as they should in the conventional circuit, and all the peculiarities of quantum-mechanical measurement are dumped into V_{REF}. We need a threshold circuit that has gain at the threshold voltage to differentiate the HIGH and the LOW levels. We discuss this issue first in Section 2.08.

The Boolean-level determination affects the measured circuit to the minimum degree: yet the circuit and the test equipment must be considered as an integrated whole. Then the test equipment does indeed disturb the tested circuit. The tested circuit after the test is in the state that provided the measured value. As for the state before the test, certain doubts remain, but this is the best that we can do by this quantum-mechanical test model.

2.08 The Decision Threshold

Boolean-level determination is carried out either by a single gate or by a latch consisting of gates or inverters. Let the gate's input voltage be V_1 and the output voltage be V_0. The capability of an inverter or a gate to make a qualitative decision at the output (V_0) on the given voltage at the input (V_1) depends on the gate's input-output characteristic specifically, how nonlinear the characteristic is. DC input-voltage and output-voltage characteristics of an inverter of Fig.2.08.1(a) are studied in this section. The result depends on the assumed FET characteristic, especially how nonlinear the FET characteristic is. The gradual-channel, low-field model used frequently before is not the extreme linear or the extreme nonlinear characteristic. Therefore, we consider the two extrema cases, using the two different FET models.

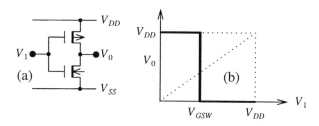

Figure 2.08.1 The V_1 - V_0 characteristic of an inverting buffer

If the FETs have the most nonlinear characteristics, they are represented by the collapsable current generators (Shoji, 1992). If $V_1 > V_{THN}$ and $0 \le V_0 \le V_{DD}$, the NFET drain current I_N is given by

$$I_N = I_{DNmax} = G_{mN}(V_1 - V_{THN}) \quad (V_0 > 0) \quad (2.08.1)$$
$$\text{but} \quad V_0 = 0 \quad (I_N < I_{DNmax})$$

The PFET drain current I_P is given similarly by

$$I_P = I_{DPmax} = G_{mP}(V_{DD} - V_1 - V_{THP}) \quad (2.08.2)$$
$$(V_0 < V_{DD})$$
$$\text{but} \quad V_0 = V_{DD} \quad (I_P < I_{DPmax})$$

Using Eqs.(2.08.1) and (2.08.2), the static equilibrium condition $I_P = I_N$ gives the switching threshold voltage V_{GSW} as

$$V_1 = V_{GSW} = \frac{G_{mP}(V_{DD} - V_{THP}) + G_{mN} V_{THN}}{G_{mN} + G_{mP}} \quad (2.08.3)$$
$$= \frac{\beta(1 - v_{THP}) + v_{THN}}{1 + \beta} V_{DD}$$

where $\beta = G_{mP}/G_{mN}$ is the beta ratio, $v_{THP} = V_{THP}/V_{DD}$, and $v_{THN} = V_{THN}/V_{DD}$. If $V_1 < V_{GSW}$, then $V_0 = V_{DD}$, and if $V_1 > V_{GSW}$, then $V_0 = 0$. The $V_0 - V_1$ characteristic is shown in Fig.2.08.1(b). The inverter has an infinite static gain at $V_1 = V_{GSW}$. If another inverter is cascaded to make a noninverting buffer, the switching threshold voltage of the cascaded buffer is the same as that of the first stage. If the inverter is used as the building block of the latch to determine the Boolean-level, the measured node voltage higher than V_{GSW} has a Boolean-level HIGH, and the voltage lower than V_{GSW} has level LOW. The Boolean-level depends on the buffer's switching-threshold voltage. The threshold inverter is a realistic observer of the node-voltage changes that is slower than the delay time of the inverter. From Eq.(2.08.3), if we do not know β of the inverter used in the latch circuit of the test equipment, we do not know the decision threshold of the latch. If the designer of the test equipment has complete freedom in selecting whatever the value of β he prefers and if the value is not disclosed, the observer does not know how the observation is carried out. This is the situation we are in at the Boolean-level determination. This interpretation is equivalent to the hidden variable interpretation of quantum mechanics: according to this interpretation there are variables in the microscopic world that are hidden, that we cannot control, and that create randomness. In the gate-field model V_{GSW} or β of the test equipment is the hidden variable.

If the FETs have the most linear characteristics, they are gate-voltage controlled linear resistors. If $V_1 > V_{THN}$, the NFET characteristic is given by

$$I_N = B_N(V_1 - V_{THN})V_0 \quad (2.08.4)$$

and the PFET characteristic is

$$I_P = B_P(V_{DD} - V_{THP} - V_1)(V_{DD} - V_0) \quad (2.08.5)$$

From the static equilibrium condition $I_P = I_N$, we obtain

$$V_0(V_1) = \quad (2.08.6)$$
$$= \frac{B_P(V_{DD} - V_{THP} - V_1)}{B_N(V_1 - V_{THN}) + B_P(V_{DD} - V_{THP} - V_1)}$$
$$= \frac{\beta(1 - v_{THP} - v_1)}{v_1 - v_{THN} + \beta(1 - v_{THP} - v_1)} V_{DD}$$

where $\beta = B_P/B_N$ is the beta ratio. If the switching-threshold voltage V_{GSW} is defined by the eigenvalue voltage of the inverter, defined by $V_0(V_{GSW}) = V_{GSW}$, V_{GSW} satisfies

$$(B_N - B_P)V_{GSW}^2 + A_1 V_{GSW} - A_0 = 0$$

where

$$A_1 = B_P(2V_{DD} - V_{THP}) - B_N V_{THN} \quad A_0 = B_P V_{DD}(V_{DD} - V_{THP})$$

Quantum-Mechanical Gate Field

The equation is solved for V_{GSW} as

$$V_{GSW} = \qquad (2.08.7)$$
$$= [\sqrt{A_1^2 + 4A_0(B_N - B_P)} - A_1]/[2(B_N - B_P)]$$

In the limit as $B_N, B_P \to B$, $V_0(V_1)$ is simplified to

$$V_0(V_1) = \frac{V_{DD} - V_{THP} - V_1}{V_{DD} - V_{THP} - V_{THN}} V_{DD} \qquad (2.08.8)$$

and it is a linear function of input voltage V_1 and

$$V_{GSW} = \frac{V_{DD} - V_{THP}}{2V_{DD} - V_{THP} - V_{THN}} V_{DD}$$

If we further assume that $V_{THN}, V_{THP} = V_{TH}$

$$V_0 = V_{DD}(V_{DD} - V_{TH} - V_1)/(V_{DD} - 2V_{TH}) \qquad (2.08.9)$$
$$V_{GSW} = V_{DD}/2$$

This is the case of a symmetrical inverter. Figure 2.08.2 shows the static characteristic for $\beta = 1/4$, 1, and 4. To swing the inverter output from V_{DD} to 0 requires an input-voltage swing from V_{THN} to $V_{DD} - V_{THP}$. The gain of the inverter is $V_{DD}/(V_{DD} - V_{THP} - V_{THN})$, and this is larger than unity but is close to it, since usually $V_{THP}, V_{THN} \ll V_{DD}$. If the FET characteristic is more nonlinear, the range of the input voltages required to complete switching becomes narrower, ultimately down to a single switching-threshold voltage V_{GSW} in the limit as the collapsable current generator. It is the device nonlinearity and circuit gain that has the capability of extracting a qualitative conclusion from a quantitative data.

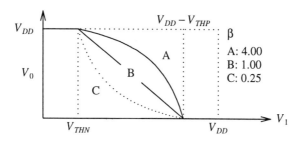

Figure 2.08.2 The V_1 - V_0 characteristic of an inverting buffer

As Eq.(2.08.3) shows, V_{GSW} is in the range

$$V_{THN} < V_{GSW} < V_{DD} - V_{THP}$$

since $0 < \beta < \infty$. Any value of β in the range is equally valid: there is no single preferred value of V_{GSW}. To set V_{GSW} at the exact halfway between V_{DD} and V_{SS} has no particular physical meanings. A complementary inverter can be pullup-pulldown symmetrical, but a multiinput gate can never be symmetrical, and its switching threshold voltage depends on the voltages of the other inputs. Whatever the definition of the logic threshold voltage may be, it must be agreed to by all the gates in the logic system, and that is impossible. There is no single and universally acceptable V_{GSW}. If V_{GSW} has no preferred value, the Boolean-level depends essentially on the test equipment.

Determination of the Boolean-level of a node voltage is a qualitative decision made from quantitative input data. Conventional thinking is that there should be no gray area between the two levels, the HIGH and the LOW. As we saw, a logic-threshold voltage cannot be set. Then how can such a black or white decision be made? Boolean-level is determined, in a practical integrated circuit testing, by the following criteria. A pair of logic-threshold voltages, V_{Hmin} and V_{Lmax}, are given such that $V_{Hmin} > V_{Lmax}$. The node voltage V is considered to have the HIGH Boolean-level if $V > V_{Hmin}$, the LOW if $V < V_{Lmax}$, and no Boolean-level if $V_{Lmax} < V < V_{Hmin}$. If V gives no Boolean-level and if the circuit-design specification requires a Boolean-level, the test result is a failure. This definition is appropriate for a conservative chip-failure screening operation. An automatic test machine has the capability of setting the two thresholds and is equipped with a complex program control to determine and interpret the test results. If the measurement is carried out by a simple logic gate, such a sophisticated capability is not available. In particular, no band of the voltage range can be set where a Boolean-level does not exist. The idea of setting up a no-man's land between the two Boolean-levels will provide, however, a basic idea further developed in Section 2.09.

Removal of this band by setting $V_{Hmin} = V_{Lmax}$ makes the definition easier, but the problems of ambiguity of the logic-threshold voltage come back. There is no way around the problem, and yet one more fundamental issue creeps in, which is related to the time required for the measurement. If the test equipment does not have an instantaneous response, there is always a question about when the measurement is carried out. If the time required by the test equipment is infinitely long, the measured data lose their meanings. A single-threshold voltage can be valid only if the voltage measurement is allowed to take unlimited measurement time in the worst case. In a realistic situation of limited measurement time, the result of determination depends somewhat on the past history of the test equipment and the circuit as well. This means that we do not know what we are measuring, much more than we ever imagined. This issue has not surfaced as a serious practical issue yet, since conventional digital circuits are still conservatively designed macroscopic objects, and we test them as if they were analog circuits. Yet the issue is fundamental, and they create a lot of confusion. We need to study a digital circuit-measurement process very thoroughly.

2.09 The Probabilistic Interpretation of Boolean Level

The Boolean-level determination carried out by each logic gate in a digital circuit is to choose either the HIGH or the LOW Boolean-level at the switching-threshold voltage V_{GSW} of the gate. In the last section we observed that the V_{GSW} of an inverter depends on the beta ratio and the conduction-threshold voltages of the FETs. If the input voltage is higher than V_{GSW}, the output voltage is lower than that, and vice versa. Generally, the output voltage is closer to either one of the Boolean-levels than the input voltage because the small-signal voltage gain of the gate as an amplifier biased at V_{GSW} is high. The V_{GSW} of a gate is set by FET characteristics that are determined by the circuit designer's choice of FET sizes. Like it or not, there must be freedom to choose V_{GSW} over a wide range to ensure circuit-design flexibility, and the Boolean-level determination must be independent of the designer's choice. The designer knows but the observer does not know, the beta ratio or V_{GSW}. There is another serious source for the switching-threshold uncertainty. A logic circuit contains gates that have more than one inputs. In a multiinput gate there is no single switching-threshold voltage like V_{GSW}. In a two-input NAND or NOR gate, the switching-threshold voltage from one input depends on the voltage level of the other input. The observer does not know the timing, either. Generally, in an N-input gate, there are N functions, each of which has $N-1$ input voltages as the arguments, that give the switching-threshold voltage of each of the N inputs. The functions are quite complex (Shoji, 1992). The only general characteristis that can be ensured is that the switching-threshold voltage is between V_{THN} and $V_{DD}-V_{THP}$. The quantization by each gate should be interpreted as being carried out at any voltage in the range used as the switching threshold. This uncertainty is essential and fundamental. The ignorance of the observer and the impossibility of setting the threshold voltage at a certain level have the same meaning as the hidden variable interpretation of quantum mechanics.

Subject to this measurement condition, what would be a rational interpretation of a Boolean-level determination experiment? The answer is to introduce a probabilistic interpretation. To cover the range of the threshold voltage, we consider that the Boolean-level determination is a probability determination that may give a definite result only some of the time.

The ambiguity of the decision threshold is an essential attribute of a Boolean-level determination of a digital circuit. In quantum mechanics similar problems exist. Is it possible to bring a measured object to an accurately specified distance from the test equipment and maintain its location for a duration of time, to generate a controlled interaction between the object and the test equipment? To do so, location and velocity of the microscopic object must be determined simultaneously. This is a violation of Heisenberg's uncertainty principle. Similarly, in a digital measurement, the switching-threshold voltage between the HIGH and the LOW Boolean-levels loses its meaning, and this leads to a probabilistic interpretation of the location of the digital excitation.

A voltage difference less than the logic amplitude is unmeasurable, and accordingly, such a voltage has no physical meaning. This leads to a crucial conclusion that the decision-threshold voltage of the latch in the test equipment can be set at *any* voltage between the HIGH and the LOW logic-level voltages with equal validity, if we proceed to model the logic circuit and the test equipment as a *macroscopic* analog circuit for our theoretical study. The measurements of the input-output characteristics of a particular gate cannot be used to determine the results of quantization, however, and therefore a particular choice of V_{GSW} is not allowed for the Boolean-level determination. Such a correlation is allowed only if all the measurements are interpreted as macroscopic analog measurements. Loss of validity of V_{GSW} is equivalent to saying that the results of digital observation becomes probabilistic. Then, does a digital circuit operate reliably and produce predictable results? It is really due to quite a few design and operational restrictions that made a logic circuit look like a macroscopic analog circuit to make it work all the time. The ratioless logic of CMOS and the positive conduction-threshold voltages of the FETs are the examples of such an engineering effort. In the ultimate limit of digital-gate scaledown the restrictions may not be enforced any more. A latch used to determine the Boolean-levels is also subjected to an intensive design effort to ensure correct operation. This is, of course, due to the latch-circuit structure, which includes inverters and logic gates as its circuit component and a clock to operate it. The decision threshold condition of a latch (which is equivalent to V_{GSW}) is more complex than a single-gate threshold V_{GSW}, and that is discussed in the next section. Here it suffices to say that the switching threshold of the latch depends on its state (more than one-node voltage) at the time when a measurement is initiated.

We cannot consider that the test equipment is outside the logic circuit: in a measurement procedure it becomes as if it were an integral part of the circuit. The functional part of the test equipment of Figs.2.07.1 and 2.07.2 is a latch. Since a latch is traditionally not considered to be a measurement circuit, a short review of its function in a digital circuit is required. Latches carry out several tasks. They are used to restore correct voltage levels representing the Boolean-levels if the levels are corrupted by loading, timing, and so on. Latches are used in the conventional synchronous logic circuit to realign the data with respect to the time frame set by its own clock. Those are really the observation of the output of the upstream logic circuit by the latch. A latch is a two-input logic gate that accepts data D and clock CK. It *captures* the input data D at the time when the clock CK makes a high to low transition. To capture the correctly processed data of a synchronized logic circuit, the clock transition time must be somewhat later than the time when the processed data arrive at the latch input. The time is called a *setup time*. A diagnostic measurement of a logic circuit is always carried out by paying full attention to the timing relation. Is such a fine adjustment of timing a prerequisite of a measurement carried out on a gate-field observation? The crucial point is that instantaneous measurement in sequence, such as monitoring the node voltage by analog test equipment, cannot be executed by the test equipment of Figs. 2.07.1 and 2.07.2. The voltage at the input node of the latch may not have reached the steady Boolean-logic level at the time of the switch closure of the circuit of Fig.2.07.1(a), that initiates a measurement. After the switch closure, the latch applies positive feedback, and after some time the output-node voltage, which is also the voltage of the measured node, settles at one of the Boolean voltage levels. This operation of latch is a complicated sensing and forcing function. It has an aspect of forecasting a future event from past and from present states. The forecasting may appear intelligent, but it is not necessarily a reliable operation. In the test equipment of Fig.2.07.1, the latch output forces the node to the Boolean-level it determined. If the latch holds the information that is consistent with the eventual outcome of the operation, the latch arrives at the conclusion before the logic-circuit node alone is able to. This operation may appear as if time skipped forward at the measured node. But on other occasions the switch is closed too early to capture the result, and the test misses the crucial development. We try to capture a digital excitation that is an undividable whole. If the tested circuit does not offer the whole, the test equipment by its own initiative must make up the missing part to complete it. If the test equipment, again by its own initiative, decides to ignore what the tested circuit presents, it discards everything. This is a decision that has a certain creative capability, and the *creativity* determines the subsequent development of the circuit. The measured result is created by the latch. It is strange that a test procedure may create something. In common sense creation is always a positive act. The creativity of test is not a positive definite attribute. In the test there can be both positive and negative *creativity*, and they compensate. This is a very important point. If observation created existence, a philosophical contention that what exists in the physical world exists only in a thinking human mind would be valid. The difficulties in comprehending the microscopic observation originate from the elementary, undividable object. In such an observation, positive and negative *creativity* balances on the probability. This is a rational interpretation that is clear from a concrete model of digital circuit test.

From this viewpoint the correct operation of a digital circuit in the conventional regime is only a small fraction of all its possible operational modes. If statistically observed, a digital circuit fails in an overwhelming majority of the cases. Logic circuits are forced to work correctly by design and operational restrictions, and this is a qualitative difference between a digital circuit and the mechanical operation of an analog circuit. An analog circuit is reliable because it is essentially a macroscopic object. A digital circuit is *made* reliable by engineering. In a certain logic circuit like CMOS, there are well-defined regions where the operation is logically correct. CMOS is the only MOS logic family that is termed as *ratioless* logic. Approximately in the order of CMOS, NMOS, and GaAs DCTL, the range of correct logic operation diminishes. NMOS and DCTL are ratioed logic.

If a node of a digital circuit is in one of the steady states, and if a latch is connected to the node to determine the Boolean-level, the latch determines the node logic level HIGH or LOW, whatever the switching-threshold voltage of the latch that is within the HIGH and LOW voltage range. Practical logic circuits are designed to satisfy this operational restriction. If the node of the circuit is in a transitory state between the two logic levels, the Boolean-level given by the latch depends on where the V_{GSW} is set. It may be appealing to set V_{GSW} at the halfway between the HIGH and the LOW voltage levels, but this preference is arbitrary. No preferred value of V_{GSW} and an uncertain Boolean-level in the transitory state are the essential features of the digital measurement, and this makes the test result probabilistic. The probability comes into the measurement since even if we know the analog node voltage, we do not know in what state the node is in, since we have no reason to select a particular logic threshold. This is the mechanism by which a probabilistic interpretation creeps into a digital circuit, very much like in quantum mechanics. Yet in quantum mechanics there is a fundamental uncertainty, and in the gate-field a similar uncertainty originates from the restricted observation, the essential mechanism of the intelligent operation of quantitative → qualitative conversion entrusted to a simple analog circuit of a logic gate. Since this is quite fundamental, let us make the crucial point clear by setting up an equivalent macroscopic analog circuit of the Boolean-level determination.

Figure 2.09.1 shows a complete schematic of a latch circuit used to determine the Boolean-level. Differential buffer A and inverting buffer B are cascaded, and switch S controls the operation. If the switch is turned on, the cascaded buffers apply positive feedback that settles the latch to one of the Boolean-levels, the HIGH or the LOW. Differential buffer A has two inputs, V_{A+} and V_{A-}, and in the narrow linear-gain region of the input voltage the output voltage V_{OUT} is given by

$$V_{OUT}(V_{A+}, V_{A-}) = \qquad (2.09.1)$$
$$= -|G|(V_{A+} - V_{A-}) + V_{OUT}(0)$$

where $|G|$ is the magnitude of the gain, which is large, and $V_{OUT}(0)$ is the output voltage when $V_{A+} = V_{A-}$ and that is within the range

$V_{SS} < V_{OUT}(0) < V_{DD}$. We assume that $|G| \gg 1$, and $V_{OUT}(0)$ is reasonably centered in the range $V_{SS} < V_{OUT}(0) < V_{DD}$. If $V_{A+} - V_{A-}$ increases beyond a limit typically $V_{DD}/|G|$, $V_{OUT} \to V_{SS}$, and if it decreases beyond the limit $-V_{DD}/|G| V_{OUT} \to V_{DD}$. V_{A+} is the voltage of the tested node, and V_{A-} is a reference voltage within the range $V_{SS} < V_{A+} < V_{DD}$.

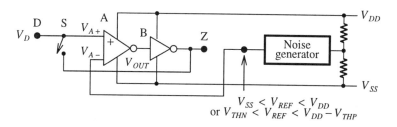

Figure 2.09.1 An equivalent circuit
of a quantum-mechanical observation

If $V_D = V_{A+}$ is either V_{DD} (HIGH logic voltage) or V_{SS} (LOW logic voltage), the latch settles at the HIGH or the LOW logic level when the switch is turned on. The Boolean-level is determined without ambiguity, since the reference voltage, V_{A-}, is in the range $V_{SS} < V_{A-} < V_{DD}$, and even if it takes a random value, the outcome of the test is definite. If V_{A+} is neither the HIGH or the LOW, V_{A-} can be higher or lower than V_{A+} with certain probability.

The equivalent circuit of Fig.2.09.1 is interpreted as follows. The differential amplifier and the buffer symbols represent ideal components that are not susceptible to microscopic effects like noise. The microscopic effects of noise and uncertainty are all pushed into the random reference voltage V_{REF} (strictly speaking, $V_{OUT}(0)$ matters as well, but if $|G| \gg 1$ the effect is negligible). In reality if the test equipment is built as an on-chip integrated circuit, the noise affects all the components and interconnects. The cumulative effects are represented by the random reference voltage. Alternatively, the random reference voltage may be considered to represent the unknown design of the test equipment, such as the β ratio. Either way, the circuit of Fig.2.09.1 should be interpreted as specified by this statement.

In the test equipment of Fig.2.09.1, we assumed that the reference voltage V_{A-} is in the range $V_{DD} > V_{A-} > V_{SS}$. Instead, if we may assume that V_{A-} is in the restricted range $V_{DD} - V_{THP} > V_{A-} > V_{THN}$, the ranges of V_D close to V_{SS} and V_{DD} are sure to belong to the definite LOW and the definite HIGH Boolean-levels, such that

$$\text{LOW:} \quad V_{THN} > V_D > 0 \qquad (2.09.2)$$

HIGH: $V_{DD} > V_D > V_{DD} - V_{THP}$

Existence of a voltage band that represents the Boolean-levels is significant, by removing some of the ambiguities of the logic level definition when the node voltage is indistinguishably close to the definite logic level. It takes infinite time for the node voltage to settle at the final V_{DD} for the HIGH and V_{SS} for the LOW logic levels. This leads to a conclusion that we never have a definite Boolean-level within finite time. This conclusion is dependent on the FET model. This margin is not necessary in the FETs modeled using the collapsable current generator model (Shoji, 1992). Although this FET model is convenient, it is an idealization. Furthermore, if a node is in either level, the node voltage has nonzero margin for loading, applied by the test equipment. Positive conduction-threshold voltage of the FETs is one of the engineering choices that make a digital circuit reliable, but it has theoretical consequences as well. The nonzero V_{THN} and V_{THP} are required to make a circuit built from realistic FETs operate properly, when the essentially quantum-mechanical features appear. The close similarity between a digital level and a quantum-mechanical state can be demonstrated by the nonzero V_{TH}, that make the digital level independent of the FET model.

The reference voltage takes random value in the range $V_{DD} - V_{THP} > V_{REF} > V_{THN}$, but should it be a static-voltage level that is unknown, or a time-dependent voltage like noise? The reason that the HIGH and the LOW Boolean-level voltages are the minimum distinguishable voltage is that the noise of the circuit influences the operation. It is appealing to assume that the voltage is a time-dependent waveform within the range, but this assumption brings a few complexities. The instantaneous noise voltage need not be limited in the range $V_{DD} - V_{THP} > V_{REF} > V_{THN}$. The noise spectrum is limited by the FET characteristics of the digital circuit, since the same FETs are used to build the test equipment. The differential amplifier of Fig.2.09.1 filters whatever high-frequency component of the noise. The actual noise spectrum depends on the test equipment. Yet we never need the details of the noise spectrum. The randomness introduced by the noisy reference or unknown time-independent reference is the same, however, as much as the result of the measurement is concerned, because the one is the ensemble average and the other is the time average. The random time-dependent reference provides more concrete physical mechanism of the problem, and the random time-independent reference voltage provides easy analysis. The model provides some idea about the level of scaledown where the quantum-mechanical effects on the circuit level begin to show up. The level of scaledown is reached when the decreasing V_{DD} becomes comparable with the noise induced on the circuit nodes by the activity of the circuit.

If V_D of Fig.2.09.1 is between V_{THN} and $V_{DD} - V_{THP}$, the Boolean-level measurement result becomes probabilistic. The probabilities of measuring the HIGH and the LOW Boolean-levels are shown schematically in Fig.2.09.2. If the voltage is less than V_{THN}, the measured result is definitely the LOW, and if the voltage is higher than V_{THP}, it is definitely the HIGH. In between the two, the probability of measuring the HIGH increases with increasing voltage, and the probability of measuring the LOW decreases. The sum of the probabilities is always unity.

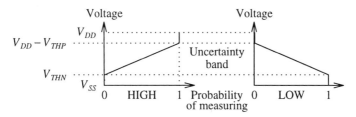

Figure 2.09.2 Probabilistic test results

The probabilistic interpretation is consistent with a peculiarity of CMOS static gates. In a multiinput CMOS static gate built using FETs whose conduction threshold voltages are positive, any node that cannot be pulled up to V_{DD}, or cannot be pulled down to V_{SS}, does not represent a Boolean function of the input digital variables. In Fig.2.09.3, if input variables A and B are settled at any of the HIGH or LOW levels, output Z will settle at HIGH or LOW levels as well.

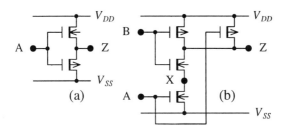

Figure 2.09.3 Nodes that do not have valid Boolean-levels

Z is a Boolean function of inputs A and B. For Fig.2.09.3(a), $Z = \overline{A}$, and for Fig.2.09.3(b), $Z = \overline{AB}$. Node X of Fig.2.09.3(b) can be pulled down to V_{SS} but can never be pulled up above $V_{DD} - V_{THN}$ if the NFET is leakageless. This means that node X can be measured at the LOW level with certainty but not at the HIGH level. Therefore, node X never represents a Boolean function of A and B, since the HIGH level is never measurable with certainty (Shoji, 1987). What can be inferred from the internal node measurement is limited to the following: if X is LOW, A is HIGH, and this is incomplete as a Boolean function. A source-follower driver cannot be used as a CMOS static-logic gate in the conventional CMOS.

If V_D is between $V_{DD} - V_{THP}$ and $V_{SS} + V_{THN}$, the level at which the latch settles depends on the value of the random-reference voltage generator of Fig.2.09.1. If V_D is close to $V_{SS} + V_{THN}$, the latch settles at the LOW with the high probability, since the random reference voltage is more likely to be higher than V_D than lower. Similarly, if V_D is close to $V_{DD} - V_{THP}$, the latch settles at the HIGH with the high probability. In a measurement using the latch of Fig.2.09.1 the probability of measuring HIGH, p_H, or LOW, p_L, can be measured but not the value of V_D. This is rational, since the measurement does not have resolution to differentiate voltage less than $V_{DD} - V_{SS}(=0) - V_{THN} - V_{THP}$. Two probabilities sum to unity

$$p_H + p_L = 1 \qquad (2.09.3)$$

since there is no other state in the binary logic circuit. Each node has probability, p_{iH} and p_{iL}. The node i voltage V_i is determined by solving the set of simultaneous circuit equations of the form

$$C_i \frac{dV_i}{dt} = I_i(V_1, V_2, \ldots, V_N, V_*) \quad (1 < i < N) \qquad (2.09.4)$$

where V_* means collectively all the input-node-voltage waveforms of the circuit. In the Boolean-level determination circuit of Fig.2.09.1, there is no reason to think that V_{REF} has preferred value in the range $V_{DD} - V_{THO} > V_{REF} > V_{THN}$. Then the probabilities are given by

$$p_{iH} = 1 \text{ and } p_{iL} = 0 \quad (V_i \geq V_{DD} - V_{THP}) \qquad (2.09.5)$$

$$p_{iH} = \frac{V_i - V_{THN}}{V_{DD} - V_{THP} - V_{THN}} \quad p_{iL} = \frac{V_{DD} - V_{THP} - V_i}{V_{DD} - V_{THP} - V_{THN}}$$

$$(V_{DD} - V_{THP} > V_i > V_{THN})$$

$$P_{iH} = 0 \text{ and } p_{iL} = 1 \quad (V_i < V_{THN})$$

Probabilities p_{iH} and p_{iL} and V_i give the same measure, but they have different physical meanings.

In Chapter 3, it is shown that the states represented by the energy eigenstates are a subset of the circuit's microstates that do not include any S-state FET (Section 3.04). The states produce minimum entropy, and they are the steady states of irreversible thermodynamics (Prigogine, 1961). The states including S-state FETs are represented by a probabilistic superposition of the microstates that do not include only the N and T state FETs. The quantum-mechanical system and digital circuit are certainly quite different objects. Yet they look quite similar in many ways, if certain idealization is made to the latter. The idealization is to pretend

Quantum-Mechanical Gate Field

not to know certain key variables. The interpretation is not valid in quantum mechanics, but a system that satisfies the assumptions does exist. The system is certainly different but still has a lot of similarity to the quantum-mechanical system. Why has the system not been sought by this time? The reason is that there is no branch of natural science that deals with the model itself. My desire is that the electronic circuit theory as the science of the models will fill the missing element.

2.10 Metastability in Observation

A digital latch used to determine the Boolean-level has a peculiar problem. If the information that has been brought into the latch by the time when the measurement is initiated is insufficient to determine the final state, the latch stays in the unsettled state for a long period of time. The effect is called a *metastability*. Because of its obvious impact on the system's operational integrity, the effect has been studied in detail since its discovery (Chaney and Molnar, 1973).

Figure 2.10.1 shows a latch whose switch S was closed to initiate Boolean-level determination. Let the initial node voltages be $V_1(0)$ and $V_0(0)$. The node voltages are determined from the set of equations

$$C_1 \frac{dV_1}{dt} = G_{mP1}(V_{DD} - V_{THP} - V_0) - G_{mN1}(V_0 - V_{THN})$$

$$C_0 \frac{dV_0}{dt} = G_{mP0}(V_{DD} - V_{THP} - V_1) - G_{mN0}(V_1 - V_{THN})$$

where the collapsable current generator model of the FETs was used and we assume $0 < V_0, V_1 < V_{DD}$. Analysis in the limited domain is sufficient because the latch generally settles in a short time if the variables get out of the domain. The set of equations is simplified to

$$C_1 \frac{dV_1}{dt} = -(G_{mN1} + G_{mP1})(V_0 - V_{GSW1})$$

$$C_0 \frac{dV_0}{dt} = (G_{mN0} + G_{mP0})(V_1 - V_{GSW0})$$

where the switching threshold voltages of each stage is given by

$$V_{GSW1} = [G_{mP1}(V_{DD} - V_{THP}) + G_{mN1} V_{THN}]/(G_{mN1} + G_{mP1})$$
$$V_{GSW0} = [G_{mP0}(V_{DD} - V_{THP}) + G_{mN0} V_{THN}]/(G_{mN0} + G_{mP0})$$

If we use $\phi_1 = V_1 - V_{GSW0}$ and $\phi_0 = V_0 - V_{GSW1}$, we get

$$\frac{d\phi_1}{dt} = -\frac{G_{mN1} + G_{mP1}}{C_1} \phi_0 \qquad (2.10.1)$$

$$\frac{d\phi_0}{dt} = -\frac{G_{mN0} + G_{mP0}}{C_0} \phi_1$$

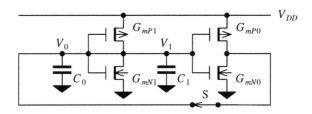

Figure 2.10.1 Metastability analysis

If we define a time constant t_A by

$$t_A{}^2 = \frac{C_0 C_1}{(G_{mN0}+G_{mP0})(G_{mN1}+G_{mP1})} \tag{2.10.2}$$

the equations are solved as

$$\phi_1 = A\exp(t/t_A)+B\exp(-t/t_A) \tag{2.10.3}$$

$$\phi_0 = -[C_1/[t_A(G_{mN1}+G_{mP1})]][A\exp(t/t_A)-B\exp(-t/t_A)]$$

where A and B are the integration constants chosen to satisfy the initial conditions. If we eliminate $\exp(t/t_A)$ and $\exp(-t/t_A)$ among the two equations, we get the two sets of solutions

$$V_1 = V_{GSW0} \pm X\sqrt{1+[(V_0-V_{GSW1})^2/Y^2]} \tag{2.10.4}$$

$$V_0 = V_{GSW1} \pm Y\sqrt{1+[(V_1-V_{GSW0})^2/X^2]}$$

where

$$Y = X\sqrt{(C_1/C_0)[(G_{mN0}+G_{mP0})/(G_{mN1}+G_{mP1})]}$$

and where X is a parameter chosen to satisfy the initial conditions.

The V_0 - V_1 relationship of Eq.(2.10.4) is shown in Fig.2.10.2. If an initial condition $[V_1(0),V_0(0)]$ is given as a point, a curve that passes the point can be found, and the later time development is determined by following the curve. The point determined from the initial condition moves in the directions indicated by the arrows, and the point directs to one of the two stable operating points $(V_{DD}, 0)$ or $(0, V_{DD})$ shown by the eye symbols.

If the initial condition $[V_1(0),V_0(0)]$ is very close to the point (V_{GSW0},V_{GSW1}) (closed circle), ϕ_1 and ϕ_0 are both close to zero at the beginning, by observing Eq.(2.10.3). If a special condition is satisfied that $A = 0$, we get

$$\phi_1 = B\exp(-t/t_A)$$

$$\phi_0 = -[C_1/[t_A(B_{mN1}+B_{mP1})]]B\exp(-t/t_A)$$

and the latch tends to the equilibrium point (V_{GSW0}, V_{GSW1}). Although this is a very special and rare condition and is susceptible to external influence, its existence cannot be ruled out. Practically, A being exactly zero would be quite unlikely. If $|A|$ is small but not zero, it takes time t_m given by

$$t_m = t_A \log(V_{DD}/|A|)$$

for a latch to settle at the final state. The settling time diverges if $|A| \to 0$ logarithmically. If this condition is approximately satisfied, the latch becomes sensitive to external perturbation, like noise. This effect is called *metastability* of a latch. This theory covers the essential physics of metastability.

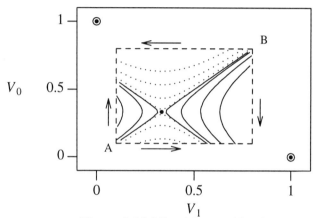

Figure 2.10.2 Decision making by latch

Metastability is still shrouded with some mystery, and the crucial point is summarized as follows:
(1) Any effort trying to eliminate metastability by inventing an *ingenious* circuit has been unsuccessful. A scenario of such a failed attempt has certain common features, such as an effort to try to remove metastability at one phase of the circuit operation creates a new metastability somewhere else, thus, the proposed solution disqualifies itself.
(2) Metastability is closely related to the fundamental attribute of an electron triode, the gain-bandwidth product limit. Then if a threshold device other than an electron triode is found, a solution may be reached. It appears impossible, however, to show the existence of a device faster than an ultimately fast electron triode, suitable for executing the threshold discrimination operation. This point is discussed in my last book (Shoji, 1996).
Points (1) and (2) become logically consistent, and the consistency is a logical barrier to further penetrating into the fundamental level.

The another important issue on metastability is the metastable state's susceptibility to minimal external influence, generally termed *noise*. If a metastable state lasts for a long time, even a small external influence determines the result of the Boolean-level determination (Shoji, 1996). In this case, there is a doubt about whether the measurement determines something physically meaningful. This question might be thought resolved if the gate-field is at absolute zero temperature. It is not. The electrical activity of the gate-field itself creates effectively a high-temperature environment to the gate-field by the electromagnetic interaction, and the effect persists.

A quantum-mechanical gedanken experiment on the microscopic level is carried out disregarding problems like metastability. Is there a fundamental difference between the two measurement procedures? Answering this question is not easy, since modeling the process of a quantum-mechanical observation cannot be as precisely specified in its details as the determination of the Boolean-level of a digital circuit. A casual observation of a quantum-mechanical experiment procedure suggests that it suffers from similar metastability problems. For instance, if an elementary particle is detected by a radiation counter, macroscopic interaction such as ionization of the molecules in the particle path is involved, which is likely to suffer from a similar effect. The crucial issue is whether it is fundamental. There are two ways to think about this problem. The first is to consider that metastability and its uncertain outcome are already included in the probabilistic result. The second is to consider the imprecisely defined details of the model of the test equipment, such as that shown in Fig.2.07.2. In the circuit model, I_P and I_N are not specified. They are the parameters that set the sensitivity or gain of the test equipment. In the limit as $I_P, I_N \to \infty$ the metastability vanishes. A quantum-mechanical measurement is to convert true microscopic information to a macroscopic information. In the process we may be assuming that the sensitivity can be infinity. The assumed infinite sensitivity is consistent with microscopic measurement in the quantum-mechanical gedanken experiment but not with the digital-circuit model. The test equipment of an elementary particle may be allowed to have infinitely sensitivity, such as a zero-delay high-gain amplifier device. In the model of the gate-field no such assumption is allowed: the observer of a digital node is a gate or a latch consisting of similar kinds of gates, and this is the assumption we must adhere to. Noise in the measurement is also a significant factor. Noise in a digital circuit is the random fluctuation originating from its own electrical activity. The noise created by electrostatic induction among the nodes is several orders of magnitude higher than fundamental noise such as thermal noise. A gedanken experiment on a gate-field that is a complex macroscopic object is carried out ultimately, including the effect, and that might be the source of the difference. These ideas are only a tentative explanation. For me, the issue has not yet been resolved satisfactorily.

2.11 Propagation of Excitation through a Nonuniform Field

The most significant parameter that determines the gate-field characteristics is the power-supply voltage V_{DD}. It does not have to be constant over the entire gate-field. Let us consider the effects of power-supply voltage variation in the gate-field. This is a prerequisite to introducing an equivalent to the most popularized effect in quantum mechanics, the tunnel effect, in a gate-field model in Section 2.12. We begin with the simple model shown in Fig.2.11.1. All the inverting buffers share the common ground, V_{SS}. The left side of the gate-field is supplied by the reduced-power-supply voltage αV_{DD}. The nondimensional parameter α is less than unity. The right side of the gate-field is biased by the full-power-supply voltage V_{DD}. In the following example, the length of the gate-field is 30, and from the location whose index equals $N_1 = 11$, the supply voltage is increased from αV_{DD} to V_{DD}. For a location between 0 and N_1, the supply voltage is less, at αV_{DD}, where $\alpha = 0.5$. The HIGH and the LOW Boolean voltages of the left side are αV_{DD} and 0, respectively. The HIGH and the LOW Boolean voltages on the right side are V_{DD} and 0, respectively. The left end of the field was grounded (the LOW level) a long time before, and a steady state is established over the entire gate-field consisting of the two different power-supply voltage zones. At time $t = 0$, the left end terminal is pulled up to αV_{DD} (the HIGH level), and a voltage stepfunction propagates along the noninverting gate-field. We study the effects of FET scaling on the transient on the right side.

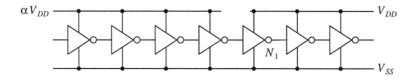

Figure 2.11.1 A gate-field having
a nonuniform power-supply voltage

At the boundary between the high and the low power-supply voltage regions, the high-voltage region inverting buffer is driven by the output signal from the low-voltage region inverting buffer, whose logic voltages are αV_{DD} (HIGH) and 0 (LOW). If the HIGH logic voltage of the low-supply voltage region is less than the inverter switching threshold voltage of the high-supply voltage region, which is given by

$$V_{GSW} = \frac{\sqrt{B_P}(V_{DD} - V_{THP}) + \sqrt{B_N}\, V_{THN}}{\sqrt{B_P} + \sqrt{B_N}}$$

the excitation from the left side is unable to proceed to the right side. At the boundary the excitation stalls. If the HIGH logic level of the left side is higher than V_{GSW}, the excitation proceeds to the right side. The numerical analysis results are shown in Fig.2.11.2. In Fig.2.11.2 V_{GSW} of the right section is decreased in the order of the figure, (a) to (d), by adjusting B_P and B_N. The potential profiles are for times 0, 25, 50, 75, 100, 125, 150, and 175. In Fig.2.11.2 only the even-numbered index node voltages are plotted to avoid clutter of the figures.

If $\alpha V_{DD} = V_{GSW}$ [Fig.2.11.2(b)], the tail of the excitation from the left penetrates into the right side, but it is never able to grow into a fully propagating excitation. If V_{GSW} is only marginally less than the HIGH logic level of the left side, the excitation is temporary stuck at the boundary, to readjust the potential profile within the field, before it starts propagating to the right side again. The crowded profiles at the boundary region of Figs.2.11.2(c) and (d) show that the excitation is temporarily stuck at the boundary.

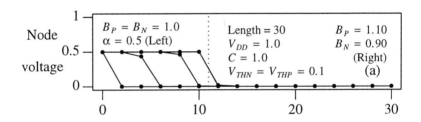

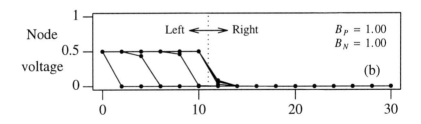

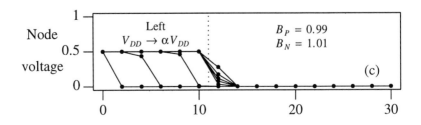

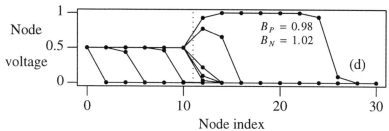

Figure 2.11.2 The propagation of excitation from a low to high V_{DD} region

As we observed before, at the boundary of the different power-supply voltages, an excitation is able to move from the low power-supply voltage section to the high-supply-voltage section, if the switching threshold voltage V_{GSW} of the high supply-voltage section is within the swing of the node voltage of the low supply-voltage section. The first several stages of the high supply-voltage section lack drive if V_{GSW} is only marginally lower than the HIGH logic voltage of the low power-supply-voltage section. The temporary stuck of the excitation at the boundary is observed from the output node waveform of the second stage of the high-voltage section. In Fig.2.11.3, the eleventh inverter is the first high-voltage section. Figure 2.11.3 shows the voltage waveform of the twelfth node.

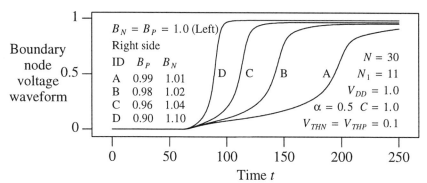

Figure 2.11.3 The transmission of excitation at a V_{DD} boundary

If the beta ratio of the high-voltage section is chosen such that V_{GSW} is less than αV_{DD}, the excitation is able to go across the boundary, but it takes extra delay time at the boundary if V_{GSW} of the high-voltage section is only marginally lower than the HIGH level of the low-voltage section. As V_{GSW} decreases, the boundary delay decreases, as is clearly observable from Fig.2.11.3. The conclusion of this section is for a classical, or analog, observation of the node voltages that does not affect the state of the node. If a quantum-mechanical observation is used, the conclusions are quite different.

2.12 The Tunnel Effect of Digital Excitation

In the nonuniform gate-field of the last section, a digital excitation from the low-voltage section does or does not continue to propagate into the high-voltage section, depending on the V_{GSW} of the high voltage section. If $V_{GSW} < \alpha V_{DD}$, the excitation continues to propagate into the high-voltage section, and if not, the excitation stops at the boundary. The excitation either propagates or stops, and there is no intermediate outcome. If the excitation continues to propagate, the digital excitation of either section carries the same information because they are both wavefronts of the propagating Boolean HIGH logic level. Then isn't it strange that the information flow stops completely at the boundary if the gate-field parameters do not satisfy the requirement? Why is the boundary so complete a barrier? Since the Boolean-levels are determined by abstracting the analog voltages out, there is a distinct group of problems originating from the elimination. The problems have the following common features: a node voltage that has a definite Boolean-level in one section may not have a definite Boolean-level in the other. Then a deterministic phenomenon in the one section becomes a probabilistic phenomenon in the other section, if the second section is subjected to an observation. This means that the information carried as the Boolean-level wavefront in the low-voltage section is able to penetrate into the high-voltage section probabilistically. An interesting aspect of this peculiarity is that the observation carried out on the gate-field at a critical location at a critical time matters essentially to determine subsequent development. The gate-field exhibiting this problem is an interesting model of both classical and quantum mechanics. In the last section we considered the classical aspect. In this section we consider the quantum-mechanical aspects. To do so, we need to consider certain fundamental issues of the observation of an elementary object like a digital excitation.

Figure 2.12.1 shows a nonuniform gate-field together with a test equipment used to determine the node's Boolean-level. This is an inverting-buffer field, shown by the circuit diagram for convenience. Change of V_{DD} takes place at the N_1-th inverter, where N_1 is an odd number. The dynamic range of the test or force equipment is from 0 to V_{DD}, since it observes the node that belongs to the high-voltage section. The left-end terminal of the gate-field is set at the LOW logic level a long time before, and it is pulled up to the HIGH logic level αV_{DD} at $t = 0$. Some time later node $(N_1 - 1)$ (even number), which was originally at the LOW logic level, pulls up to the HIGH logic level, αV_{DD}. If the high logic level is less than the switching-threshold voltage V_{GSW} of the high-voltage section, the excitation is unable to propagate to the right side, as we observed in the last section. That is the gate-field operation governed by the classical mechanism. In the following quantum-mechanical test, the parameter values of the high-voltage section are chosen such that no *classical* excitation propagation takes place. The test equipment determines the Boolean-level at node N_1. The schematic of the test equipment is shown enclosed in the dotted box of Fig.2.12.1.

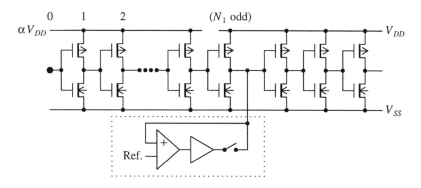

Figure 2.12.1 The tunnel effect of excitation

The reference voltage, to which the node N_1 voltage is compared, is between V_{THN} and $V_{DD} - V_{THP}$, but exactly where the voltage is cannot be stated because of the quantum-mechanical nature of the measurement. As the node $(N_1 - 1)$ voltage approaches the HIGH level, the node N_1 voltage decreases from V_{DD} but not down to V_{GSW}. The reference voltage of the test equipment, V_{REF}, is unknown, and it can be higher than V_{GSW}, and therefore the decreasing node N_1 voltage may become less than V_{REF}. If the switch of the test equipment is turned on under such a condition, the test equipment draws current from node N_1, thereby completing the pulldown. If this process takes place, the excitation from the left side continues to propagate to the high-voltage section. Propagation of excitation from the low- to the high-voltage section, however, does not take place every time when the Boolean-level determination is carried out. Since the reference voltage V_{REF} of the test equipment is random within the range from V_{THN} to $V_{DD} - V_{THP}$, we may define only the *probability* of transmission from left to right. If pulldown of node N_1 from V_{DD} is not significant, the probability of V_{REF} being higher than the node voltage is small, and the transmission takes place rarely. If V_{GSW} of the high-voltage section is only slightly higher than the HIGH logic voltage of the low-voltage section, and if the inverters in the high-voltage section are approximately pullup-pulldown symmetrical, the transmission takes place at about 50% probability. From this observation, the essential feature of a quantum-mechanical test is to feed the comparator with a random voltage within the range. In a conventional operation and testing of a digital circuit, the reference voltage is set at the selected level and held unchanged, to make all the test referenced to the same logic-threshold voltage. This is an *engineering* choice not necessarily consistent with the real nature of the digital circuit. Figure 2.12.2 shows the results of a numerical analysis.

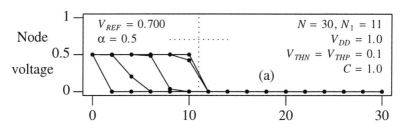

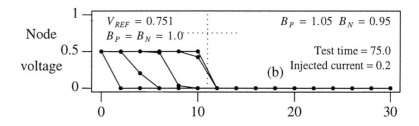

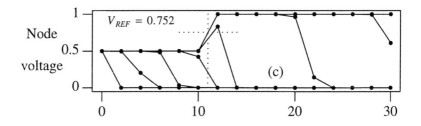

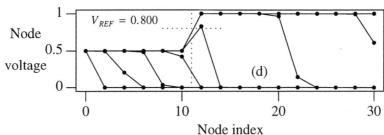

Figure 2.12.2 The tunnel effect of excitation

The gate-field has thirty cascaded CMOS inverters, and the inverters numbered 11 to 30 are powered by the full power supply voltage $V_{DD} = 1.0$. Inverters 1 to 10 have the half-power supply voltage,

$\alpha V_{DD} = 0.5$. The other parameter values are given in the figure. Only the voltages of the even-numbered index nodes are plotted to avoid clutter of the figures. The low-voltage section of the gate-field is pullup-pulldown symmetrical. The high-voltage section has $B_P > B_N$, and the threshold voltage V_{GSW} is

$$V_{GSW} = \frac{\sqrt{1.05} \times (1-0.1) + \sqrt{0.95} \times 0.1}{\sqrt{1.05} + \sqrt{0.95}} = 0.51$$

which is only slightly higher than the HIGH level of the low-voltage section, but enough to stop propagation of excitation if no observation is carried out. The reference voltage of the test equipment, V_{REF}, is random but constant during the test and is changed from 0.7 to 0.8. The test equipment details were shown earlier, in Fig.2.07.2. The test equipment injects or extracts current from node N_1 by the rule

$$I = I_0 \quad (V_{N1} > V_{REF}) \quad I = -I_0 \quad (V_{N1} < V_{REF})$$

where I is positive if the current is injected to the node. The current generator collapses if the absolute value of the current is less than I_0. In the numerical analysis $I_0 = 0.2$ was used. One more condition, the time when the switch of the test equipment is turned on to initiate the test, was set at $t = 75$. At about $t = 65$, the upgoing signal arrives at node N_1, and therefore the test is initiated soon after that. As V_{REF} is changed from 0.7 to 0.8, the gate-field changes from the stopping to the transmitting mode. The critical-reference voltage V_{REF} equals the minimum-node N_1 voltage reachable by the high-voltage section inverter that is driven by input voltage αV_{DD}, the HIGH level voltage of the low-voltage section. The minimum is about 0.75. If V_{REF} is less than that, the excitation is stopped at the boundary, as is shown by Figs.2.12.2(a) and (b). If it is higher than that, the excitation proceeds to the high-voltage section, as is shown by Figs.2.12.2(c) and (d). In this model if no test is carried out, the excitation is always stopped at the boundary. In this sense the observation created the traveling excitation in the high-voltage section.

Whether an excitation can go across the boundary depends on the time when the test is initiated, and this is the second crucial parameter selection of the test. Since the test equipment is a latch, if it once settles to a state, then the state is held by positive feedback. If the test is initiated before the wavefront arrives at the end of the low-voltage section, the latch settles at the HIGH logic level, and when the excitation arrives later, it is unable to perturb the level, since it is held by the test equipment that has a strong drive capability. In Fig.2.12.3, V_{REF} is set at 0.8, a high enough level to capture and transmit the excitation from the low-voltage section. The excitation arrives at the boundary at about $t = 66$. If the test is initiated too early, as in Fig.2.12.3(a) and (b), the excitation stops at the boundary. If the test is initiated later, the test equipment detects excitation, determines that the Boolean-level of node N_1 is LOW, and relays the information to the high-voltage section.

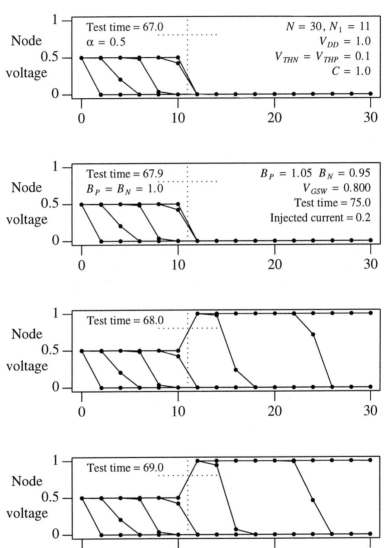

Figure 2.12.3 The tunnel effect of excitation

The two examples of Figs.2.12.2 and 2.12.3 show that a test is a critical step to determine transient development in the high-voltage section. In the two examples, V_{REF} independent of time was used for simplicity. The reference voltage is uncertain, and that makes the tunnel effect a probabilistic phenomena.

Figure 2.12.4 shows a standard speedup technique of a heavily capacitively loaded circuit node N like a memory readout circuit. The circuit, including a differential amplifier and the current driver in a feedback loop, is called a *sense amplifier*. The current generator controlled by the differential amplifier injects or extracts current from the node, thereby settling the node to the final voltage level rapidly. Figure 2.12.4(b) shows the node-voltage waveform schematically. The dashed curve shows the node-voltage waveform if the sense amplifier is deactivated. It takes time t_0 for the node to reach the logic threshold, chosen here at $V_{DD}/2$. If the reference voltage of the differential amplifier is set higher than the threshold, the node waveform is as shown by the solid curve. The switchover point moves from A to B. We note, however, that the sense amplifier must be activated at the right moment: if it is turned on too early, the amplifier adds to the load and slows down the transient or prevents switching altogether. The circuit is, in its operational principle, a quantum-mechanical test equipment except that the V_{REF} is set and held at the preferred level to expedite switching.

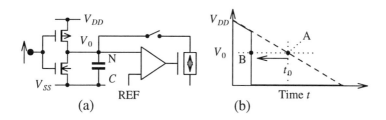

Figure 2.12.4 A sense amplifier as a tunnel-effect device

Detailed analysis of the tunnel-effect model of a gate-field showed the bizarre nature of observation on an elementary object very clearly by a simple circuit model. The problems of observation are so fundamental that we need to critically evaluate what we learned from the tunnel-effect model. The two objectives of physical observation to gain information are to
(1) Prepare the state of the object or set the object to a definite initial state before it goes through the subsequent experimental process
(2) Find the state of the object that has gone through the development and was affected by that
Both (1) and (2) are required to compare the theoretical deduction with the experimental result. In an electronics experiment, the importance of (1) is often not recognized. The input signal is considered *forced* by a signal source, and confirmation of the input state is not considered necessary. This interpretation is valid only for a classical or a macroscopic object like conventional circuit. In a quantum-mechanical object, no

theoretical conclusion can be drawn without first setting the object's state and knowing what it is by (1). The information about the initial state is missing without it.

In (1), the state of the object immediately after going through the observation is the the state determined by the observation. In (2) the state determined by the experiment is considered to be the state of the object before the observation. In this case the state after the observation is again definite, but it does not matter. As we discussed in Section 2.09, we get only a probabilistic result. Thus, (1) and (2) are significantly different observations. In a classical observation, both (1) and (2) provide definite results, and the observation determines the state of the object that is not influenced by the observation process in any way. We need to consider two issues: (a) which of (1) and (2) is a more fundamental method of observation, and (b) what is done by the test equipment we use.

The quantum-mechanical Boolean-level determination we consider in this Chapter carries out test procedure (1). The test produces either the HIGH or the LOW Boolean-level, and the node is set at the level after the test is over. Because the state after the test is guaranteed, this is test procedure (1). As for test (2) the state before the test, the test produces either the HIGH or the LOW level, and if the same test is repeated on the same circuit for many times, each time the tested circuit is prepared by test (1), the probability of measuring HIGH or LOW can be determined. Metaphorically, the test is sure about the state after but is not sure about the state before. The asymmetry is due to the impossibility of keeping the tested circuit undisturbed by the test, the impossibility of knowing the decision threshold, and the built-in capability of setting the state after the test. Since certain energy must be drawn from the circuit, this asymmetry cannot be avoided because of the causal relationship. From this observation, test (1) is more fundamental than test (2), (1) is the quantum-mechanical test, and if (2) is always deterministic, it may be called the classical test. Test (2) cannot be deterministic if applied to a microscopic object. The confusing issues of digital circuit theory originate from allowing test procedure (2) into a circuit that accepts, because of its simple nature, only test (1).

2.13 Ambiguity in the Cause and Effect Relationship

The tunnel effect of digital signal in the gate-field shows most clearly that the answer of a logic operation is the product of the act of observation. The physical phenomena, the electrical transient in the gate-field caused by the input signal to the field, affect the measurement result often only probabilistically and, only in certain special cases, deterministically. Then why does the fundamental contention of physics, causality, a given cause producing a definite, predictable effect appear to be violated? This is the topic discussed in quantum mechanics, and the

Quantum-Mechanical Gate Field 131

argument is often unconvincing from the classical physics viewpoint. In our gate-field model it is possible to reach the bottom of this question because the gate-field model is precisely defined by classical physics.

The reason that the cause and effect relationship appears violated is information compression: measuring the classical analog circuit gains much more information than measuring the quantum digital circuit. The object of measurement the node voltage profile of the gate-field has spatial extension, spreading over several gate stages. If the digital circuit is observed at a node, the voltage change does not occur instantly. The logic-circuit theory discussed in Section 1.08 removed the problems associated with the spatial domain and time domain spreading by assuming crisp switching waveforms. If that is not allowed, the location of the information-carrying object in the gate-field cannot be determined beyond the size limit. Subject to this condition, if the direction of motion of the object must be determined, we need to depend on something other than the conventional digital-node observation method.

Consideration of this point is not just academic curiosity. How can the fundamentally probabilistic gate-field (or generally a digital-circuit) operation be used to carry out a highly deterministic logic (or generally mathematically predictable) system operation? This question sounds quite remarkable, since digital circuits are generally considered more reliable in a hostile (noisy) operating environment. The answer is that the conventional digital circuit is a product of sophisticated engineering, whose intention is to hide the digital circuit's essential quantum-mechanical nature.

The reason that the most fundamental quantum-mechanical Boolean-level determination becomes probabilistic is that the logic-threshold voltage cannot be set at a definite level (certainly not outside but not even at some level within the Boolean HIGH and the LOW voltage range). This is not the problem of the observed gate-field or of the digital circuit, but it is the problem of the observer, which has intrinsic variability in the judgment standard. The demand of logic-design flexibility to allow any logic operation, by cascading any conventional gate in any way, demands that variability.

If we set the logic-threshold voltage at a certain level, we do indeed get a definite number for a digital-circuit delay time. Let us call this scheme the *classical delay determination*. The problem of the classical delay determination is most clearly observable in the difficulties discussed in Section 1.11. Here we study the same problem from the viewpoint of the cause and effect relationship. Let us reiterate this point by studying the example appeared in Section 1.11.

By using a classical method of observation of digital excitation that is, by setting a logic-threshold voltage a rational cause and effect relationship cannot be established in general cases. This can be understood from the example. Figure 2.13.1 shows the node waveforms of the three-stage cascaded inverters. The first stage A pulls down quite slowly. The second stage B has a high switching-threshold voltage, and the node pulls up fast. The third stage C has a low switching threshold, and it pulls down fast. The parameter values are indicated in the figure. We have no reason to claim that this parameter choice is irrational. Suppose that the excitation propagation is observed by setting $V_{DD}/2$ as the logic-threshold voltage as shown by the dotted line. Then in this inverter field the signal appears to propagate backward, from the third, to the second, and to the first inverter outputs. If the level is set at $0.9V_{DD}$, the excitation moves from node A to C and then back to B. This is an inevitable consequence of the classical observation of the location of the excitation in the gate-field. This conclusion was reached by an improper classical observation. The cause and effect relationship contradicts the conclusion derived from the structure of the field. Gate B receives the signal from gate A, and gate C receives signal from gate B. Classical measurement creates confusion in the cause and effect relationship.

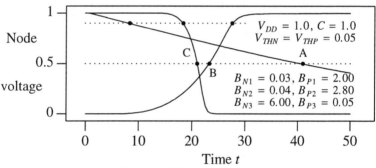

Figure 2.13.1 An apparent violation of causality in the tunnel effect

Let us reexamine this conclusion from a basic viewpoint. From the structure of the cascaded inverters A, B, and C, in that order, we conclude that the signal must propagate in that order. This is, however, extraneous information that should not bias the interpretation of the limited observational results of the gate-field, the node voltage waveform by the Boolean-level test circuit of Section 2.12. The issue is that a gate-field observer has no apriori reason to suspect the order determined by the observation. The gate-field is within a black box whose inside the observer is unable to examine. From this viewpoint it is perfectly legitimate that the observer considers that the order of connection is C, B, and A, by the observation of node waveform of Fig.2.13.1 at the logic threshold $V_{DD}/2$.

This point is consistent with the complementary relationship between the location of digital excitation and its velocity, as is discussed in Section 1.04. The location and velocity of excitation cannot be determined beyond the limit set by Planck's constant. In an extreme case such as that shown in Fig.2.13.1, the Planck's constant should become large, and if the arrival time of the excitation is determined precisely, the location of the excitation becomes ambiguous.

It may be considered that if a precision analog measurement is allowed, the node that begins to change first, A, is the first node, node B begins to change next, and node C is the third. This argument is inconsistent with the quantum-mechanical measurement method. Voltage levels between the Boolean HIGH and LOW levels can be determined only probabilistically, and analog measurement of this sort cannot be carried out on the idealized digital circuit by quantum-mechanical test equipment. Quantum-mechanical observation of this process is carried out by connecting the test equipment of Fig.2.09.1 to each node. The equipment determines the probabilities of measuring the HIGH or the LOW levels. Since the result is *not* a definite answer but only a probability, it is impossible to determine the cause and effect relationship this way: it does not make sense to say that an event is the *cause* of the effect with certain probability. This observation leads us to an important conclusion that the delay time of a gate is valid only as a classical parameter that loses validity in the quantum-mechanical modeling of the gate and this is a fundamental issue.

Information compression confuses the cause and effect relationship in another way. Unless a single input signal can be well separated from the others, often it is impossible to identify it as the cause. Let us elaborate these statements by examples. The following examples involve more than one excitation. Figure 2.13.2(a) shows one location in a gate-field occupied by an AND gate. Suppose that the gate is driven by two signals A and B, shown in Fig.2.13.2(b). A is a slow-rising signal that starts before B. B is a fast-rising signal that starts after A and that reaches the final steady-state level before A. Which of A or B is the cause of output C transition? There are no criteria for answering this question. Even if one input waveform started to change substantially earlier than the other, we cannot confidently say that the later-arriving signal created the node C transition. The cause and effect relationship becomes fundamentally undecidable. This issue has practical significance in defining the critical-path delay time in a gate-level delay simulator used to verify a design of a VLSI chip.

Suppose now that signal A is kept at the steady, HIGH Boolean-level to maintain the AND gate enabled and signal B makes a low to high transition. The cause of the output C transition can be identified as B. This conclusion is reached, however, not by simple computerized logical reasoning. There are two ways to reach the conclusion. One is to rely on a preexisting *knowledge* that an enabled AND gate is logically equivalent to a noninverting buffer. This knowledge was gained first by human minds, and was transferred to the computer. The other way is based on observation and the assumption that (1) no activity took place at node A and a switching took place at node B, (2) switching of node C took place after switching of node B, and (3) the AND gate is unable to produce any spontaneous transition internally. In classical observation, (1) requires observation over a period of time, (2) requires intelligent interpretation, and (3) is based on preexisting knowledge. All three ways depend heavily on existing knowledge, a complex sequence of reasoning, and an intelligent interpretation of the observed data.

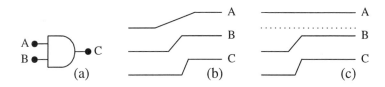

Figure 2.13.2 The ambiguity of the cause and effect relationship of interacting excitations

A multiple-input gate in a gate-field works as a location where two or more excitations traveling down the field collide and create collision products. The process of collision of particles cannot be described deterministically, even within the framework of classical mechanics. There are several possible outcomes: the collided particles may fuse together, they make elastic collision and fly away intact while maintaining each particle's identity, or they may break into more than two particles. Only the second case can be handled by classical mechanics alone. On the level of elementary particles, the number of particles changes most of the time, often an unstable intermediate product appears, and then it decays to the many stable products. These elementary particle phenomena have their parallels in the gate-field phenomena. Hazard, discussed in Section 2.16, is equivalent to a collision of two elementary particles ending up either in pair annihilation or in unstable daughter-particle creation. Collision involves physics at a level different from the mechanics of the individual particles, either classical or quantum. In the classical collision problem, knowledge of the particle as a whole, generally at the level of chemistry, is required. In the quantum collision, knowledge of the basic nuclear forces is required. In the gate-field, the collision can be handled within the framework of circuit theory, and this is one aspect of circuit theory being self-contained and fundamental.

The examples of quantum mechanics provide some insight into the ambiguity of the cause and effect relationship in the gate-field. The relationship can be established only by a *single* cause. Two particles colliding head on are not considered to have two causes but make a joint single cause. A cause cannot be separated into many causes, since cause itself is an elementary object of investigation. If this undividable object is conceptually divided, which cause is the *real* cause becomes a debatable issue. This is similar to the process of Boolean-level quantization. If an unknown logic-threshold voltage is inserted to separate the already undividable Boolean HIGH and LOW voltage levels, the results of the Boolean-level determination become probabilistic. In the example of the two-input AND gate, if input A is kept at the HIGH level that enables the gate and the input B switches, the probability of A being the cause is zero, and the probability of B being the cause is unity. Although the definition statement is simple and rational, it is not obvious how the probability should be computed.

Based on the two examples related to cause and effect relationships in digital circuits, we come to the following conclusion. The only valid delay-time measurement is to use the time-measurement setup integrated into the measured circuit. Then we may hope for a definite, unambiguous answer. Two time-measurement setups, one to introduce the signal and the other to measure the signal arrival time are required. The measured circuit is connected between the two setups. The time measurements are carried out by latches. The time-measurement method is relevant to the engineering sophistication to convert the essentially quantum-mechanical digital circuit to a semiclassical object that provides the definite answer to the given input signal. The most elementary logic circuits involve the two banks of latches as shown in Fig.2.13.3. The clocks are the cause, and the data captured by the master latches are the effects. We note here both the slave and master latch clocks are the causes: the quantum reality is created by the measurement by the master latch, when it captures the processed Boolean-level. This issue is discussed in the following sections.

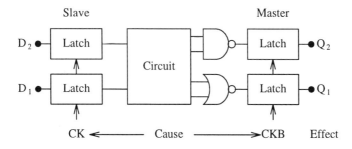

Figure 2.13.3 A cause and effect relationship

2.14 Valid Delay-Time Measurement of the Digital Circuit

In the preceding sections we developed the model and method of quantum-mechanical delay-time measurement. It may be exotic to digital-circuit engineers. The exercise was necessary to demonstrate the fundamental problems of delay-time measurement. In this section we go back to current engineering practice and explain its meanings from the new viewpoint. Defining the delay time of a logic gate (or a gate chain) requires the following steps: feeding a digital signal to the gate chain, record the signal's entry time, and then determine the time of arrival of the processed signal at the destination. For the measurement procedure to make sense, the Boolean state at the output must reverse from what it had been. The entry-time setting and the arrival-time determination are both carried out by the latches. Latches can switch the digital state (thereby providing the input signal level at a definite time) and quantize (or in its classical equivalent, interpret) the output signal by accepting the processed data. The destination latch captures the signal from the gate or the gate chain subjected to the test. The time elapsed between the entry and the arrival at the respective latches can be determined by focusing attention on the latch operation. Obviously, the two latches must be similar, so that the time references are the same. Although certain fundamental uncertainty still remains, the latch-to-latch delay time is more clearly determined than the classical gate-delay time, whose determination relies on the logic-threshold voltage. This delay-time measurement is essentially a quantum-mechanical measurement, and it is the opposite extremum of the classical delay-time measurement using the logic-threshold voltage. Not only is the method fundamentally different, but the method is more accurate and realistic in describing the real digital-circuit operation, than the classical method is. The restrictions set on the latches used in this measurement are the result of compromisation, or engineering design as in the conventional digital circuit that is an object simulating a classical, deterministic object, although its real nature is quantum-mechanical. By the definition, the latch-to-latch delay time is not a sum of the delay times of the gates in the chain, measured by the classical method.

The latches are the key elements of the quantum-mechanical delay-time measurements. Let us define the simplest structure of the latch, shown in Fig.2.14.1. We assume a CMOS implementation, so the switches are really transmission gates, but only the essential functions are shown by the simple switches. Although a real transmission gate has delay, that is ignored. The delay time of the switch is included as a part of the buffer delay time. Then the switches are assumed to ON/OFF instantly. Figure 2.14.1 shows the latch configuration that accepts a digital signal presented at input D. If switch A is open and switch B is closed, the latch retains the previously captured data. This is an idealization to bring the essential point upfront. Subject to the condition for the

Quantum-Mechanical Gate Field

latch to work, the buffers must have nonzero delay time. An interesting point is that if the buffer delay time is precisely zero, latch behavior becomes unpredictable. This is a subtle point that requires interpretation. Whatever the criteria of the state of the latch may be, nodes Q and $\overline{Q}$ take the same and the complementary Boolean-level of D, respectively, if the buffer delays are zero. Then the latch adds nothing to the input data D: the latch is equivalent to a two-stage cascaded noninverting buffer that reproduces the input signal without delay, when it is accepting signal D (switch A closed, B open). Nonzero delay time of the buffers means that the information introduced into the latch is some form of average of the past waveform, typically over the time period of the sum of the delay times of the buffers. The latch's essential function is to integrate the input signal over time. This averaging is not an equal-weight average: since signal D reaches $\overline{Q}$ first and then Q later, but Q and $\overline{Q}$ are equivalent after the latch is closed (switch A is open and B closed), the more recent voltage level at D is reflected more strongly to the final state of the latch after the switch position changes. This mode of operation is crucial in creating intelligent judgment about the signal. Operation of the latch assuming zero buffer delay is often considered as the idealized latch operation, but this is only for low clock frequency, where the logic signal settles at the final level well before the latch captures data. In this mode of operation, the latch is opened, waits for all the sputious transients to pass, and, as the final logic level is established, closes the input. It just holds the data. One significant difference between a low-speed and a high-speed circuit is the role of the latches and how much intelligent capability is demanded.

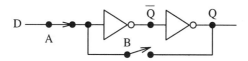

Figure 2.14.1 The structure of a digital latch
for measuring delay time

The second function of a latch is to maintain the Boolean-level representing the last input data while the processing is going on in the downstream logic chain. This capability is required because the voltage waveform representing logic data are a stepfunction and have only one transition edge per clock cycle, at the most. The stepfunction-based signal processing is consistent with the conventional static-logic circuit concept that generates the answer in an asymptotic state as time tends to infinity or the output level after an infinitely long time is always the correct answer. A stepfunction contains an infinitesimally low-frequency

Fourier component because of the requirement. The asymptotic settling to the final level is terminated at some point by capturing the processed data and by reintroducing new input data to carry out the next step processing. A latch has the capability of setting up data blocks in a continuous flow of time, thereby allowing sequence of processing. To make the processing load to fit into the limited time interval, a pair of latches carries out the function of logic-circuit delay-time measurement, and that is quite a natural role. Thus, a latch has three extra functions (1) maintaining the data format as a stepfunction that occurs only once in the clock period, (2) chopping the continuous flow of time into pieces and setting the time of initiation of new cycle, and (3) measuring the time required to process data. The synchronous sequential logic circuit schematically shown in Fig.2.14.2 executes the three functions. The latch's role in digital circuit is never simple.

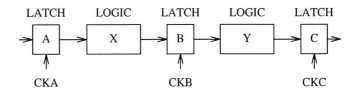

Figure 2.14.2 A synchronized sequential logic circuit and delay measurement

Instead of the truly quantum-mechanical measurement circuit discussed in Section 2.09, digital circuits use the latch of Fig.2.14.1 to introduce the signal and then to determine the Boolean-level at the output. Then, is there any convenient and consistent definition of the time when the signal has passed through the latch circuit? The following definition is the rational one. Suppose that switch A is open, switch B is closed, and the cross-coupled inverters apply positive feedback to settle at the state having definite Boolean-level and retain the one bit of information. There are two stable states, in the first one $\bar{Q}$ is LOW and Q is HIGH, and in the other $\bar{Q}$ is HIGH and Q is LOW. Suppose that the latch was in the first state and the input node D is a constant LOW logic level. In this state switch A closes, and simultaneously switch B opens. Then node $\bar{Q}$ voltage goes up, and node Q voltage goes down. After some time switch A opens, and switch B closes. Then the latch may reverse to the first state if $\bar{Q}$ has not been pulled up significantly, and it may proceed to the second state, if $\bar{Q}$ voltage has been pulled up close to the power-supply voltage. Obviously, there is a boundary between the two regimes, depending on the time when the switch A opens and switch B closes. The critical time when the switch transition changes the final state from the one to the other is the time when the latch supplied the input signal to the output node Q. We note here that node Q is loaded by the gate chain, and that affects the critical time determination. This issue is important, and it is discussed in the next section.

This definition can be used to determine the time when the latch captured the data from the logic circuit. The output of a logic gate chain drives the input of the destination latch, if switch A is closed and switch B is open. At some later time switch A opens and switch B closes. If the input voltage to the latch was going up, node Q will take HIGH logic level after some time, and if the input voltage of the latch is going down, node Q will take LOW logic level after some time. Then switch A opens, and switch B closes. The latch may proceed in the same direction after the switch-position change, or it may reverse the direction. The critical boundary between the two regimes is the time when the latch gets the processed data. This definition is that an overall judgment or *interpretation* of the input signal by the latch circuit is necessary to determine its state. Specifically, if the input voltage waveform is not monotonous, the intererpretation is quite complex. Yet the definition is simple, since only the latches are crucial, and their characterzation is not complicated. This method determines the time of entry and exit of the signal to the latch, up to the precision where the metastability becomes the issue (Section 2.10). The effects of metastability cannot be removed, and the time definition has at least that much ambiguous. Yet this precision is as high as we can hope for.

An interesting aspect of this measurement method is that if the logistics of the delay-time measurement procedure is considered, the measurement requires many trials to find the boundary between the two final states of the latch. There is no way to determine the critical time by a single measurement: the measurement is essentially an *ensemble* observation. This is a very curious and yet fundamental aspect: a quantum-mechanical delay-time measurement requires many measurements, and the classical delay-time measurement requires only one trial. This is another consequence of the quantum-mechanical measurement. The latch measures an *expectation value*, a single number, instead of the classical delay time, that is also a single number. Then the probability distribution must be determined, so ensemble measurement becomes necessary, anyway. Correct logic circuit operation requires that the measurement is less than a maximum.

The Q node of the destination latch is also loaded by the circuit, which uses the digital signal captured by the latch. The loading affects the critical time of capture of the processed signal. There are two arbitraries in the logic-chain-delay measurement, the loading capabilities of the source and the destination latches. How do we choose proper loading of the latches to get a meaningful delay time? The arbitrariness can be eliminated by considering the problem as follows. Although it is simple and cumbersome to use, a latch is an equivalent of a clock: it is a time-measurement equipment. Then certain parameters of the two latches must be kept identical to keep them synchronous, and certain other parameters need not be identical to keep them synchronous. The relative capacitive loading capability and the relative capacitive impedance looking into the latch's input are the parameters that must be kept identical. This is because the former determines the delay time required to deliver signal to the downstream logic circuit, and the latter determines the time required to feed a signal into the test circuit. The scale factor of the latch must be adjusted so that the two parameters become identical. The two latches need not have the same size, however. Rather, if the sizes of the FETs used in a latch are scaled by a constant factor, the original and the scaled latches have same characteristics in a similarly scaled circuit environment. To say this simply, a valid delay-time measurement can be carried out if the source and the destination latches find themselves in the same environment. If the two latches have different FET size ratios, the two latches have a built-in difference in the setting of the origin of time, and a measurement based on such a latch pair is considered unacceptable from the fundamental theoretical viewpoint (although that is done in practical circuits to take advantage). This adjustment removes the ambiguity in the time measurement.

The pair of definitions of the time when the signal passed through the latch gives a rational definition of the delay time of a gate chain that is inserted between the two latches. This definition is a compromise between the quantum-mechanical and the classical definitions of the delay time, but it is much closer to the quantum-mechanical definition than to the classical definition. It is not an exact quantum-mechanical definition because the measured object, the gate chain's output node is not influenced by the measurement. The switch A opens and isolates the input circuit from the latch. The destination latch's influence on the logic circuit is removed, and that is, strictly speaking, not quantum-mechanical. The tested results are not *created* by the latch as they were in purely quantum-mechanical measurement.

The difference between the quantum-mechanical Boolean-level determination and the method using the isolated latches described in this section is that here the input node of the latch is not forced to the determined Boolean-level. This is because a logic circuit is still a macroscopic object that can be modified to meet the system's requirements. Suppose that the logic chain is long that the middle point of the chain and the end point of the chain are measured by the quantum-mechanical method. Then the middle-point measurement may affect the end-point measurement, if the latter takes place some time later than the former. This is obviously not convenient for designing the logic circuit in general. Such a problem occurs because a logic circuit is a product of elaborate engineering that intends to wipe out the quantum-mechanical uncertainty.

It is interesting to note that the compromising nature of the delay measurement suggests a useful circuit response speedup method. The circuit shown in Fig.2.14.3 has a slow node such as a bus, N_C. If a quantum-mechanical measurement is carried out at the node, the response of the circuit may speed up. The inverter A has a low V_{GSW}, and inverter B has a strong current-drive capability. Then the strong node pullup begins soon after the node voltage reaches the low V_{GSW} of inverter A. The pullup expedites the signal transmission through the logic chain. This circuit, however, assumes that the signal expected at node N_C is an upgoing stepfunction front. For the reverse front the response is delayed. In a logic circuit we often know which polarity front comes, so it is possible to play a game like this.

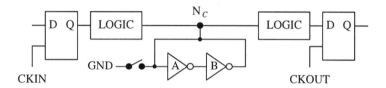

Figure 2.14.3 Quantum-mechanical measurement for node-response speedup

The operation of a latch can be reinterpreted as follows. I suggested that a latch has the capability of interpreting the output waveform of a logic chain that may be nonmonotonous. An important point is that this capability is acquired only if the latch has a propagation delay. A latch that does not have a delay cannot do the interpretation. The latch is a cascaded even number of gates that is able to store the information contained in the input waveform. When the latch is set into positive feedback, the latch makes an *executive* decision, depending on the two or more internal node voltages.

2.15 The Quantum-Mechanical Delay Definition

We continue to discuss how the quantum-mechanical measurement method is used in digital circuits. Let us see how quantum-mechanical delay time is measured in a digital circuit by a real example. Figure 2.15.1 shows the circuits used in the delay-time measurement. The delay of the inverter shown by the dark triangle (inverter 3) is measured. Inverter 0 is the signal source. Inverters 1 and 2, together with a switch, make the latch that introduce the digital signal to the tested inverter 3. Inverters 4 and 5, together with a switch, make the latch that determines the arrival time of the processed signal. The switches are assumed ideal, in that it ONs and OFFs instantly, and there is zero and infinite impedance in the ON and OFF states, respectively. This assumption is not unrealistic for modeling the latch response, if proper capacitance loading is made to the nodes of the latch inverters.

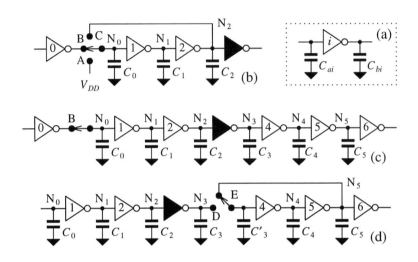

Figure 2.15.1 Quantum-mechanical delay definition and measurement

Suppose that each inverter 0 to 5 has capacitance C_{ai} looking into the inverter input and capacitance C_{bi} looking into the inverter output, as shown in Fig.2.15.1(a), (where suffix i is the inverter index). From C_{ai}s and C_{bi}s the node capacitances C_0 through C_5 are determined, depending on the switch configuration. In the circuit of Fig.2.15.1(b) with switch at B position,

$$C_0 = C_{b0} + C_{a1} \quad C_1 = C_{b1} + C_{a2} \quad C_2 = C_{b2} + C_{a3}$$

If the switch is in position C,

$$C_1 = C_{b1} + C_{a2} \quad C_0 + C_2 = C_{a1} + C_{b2} + C_{a3}$$

where the jointed node $N_0 - N_2$ includes input capacitance of inverter 1, output capacitance of inverter 2, and input capacitance of inverter 3. Similarly in the circuit of Fig.2.15.1(c)

$$C_i = C_{bi} + C_{a(i+1)} \quad i = 0, 1, 2, 3, 4, 5$$

and in the circuit of Fig.2.15.1(d) with the switch in position E

$$C_i = C_{bi} + C_{a(i+1)} \quad i = 0, 1, 2$$

and

$$C_3 = C_{b3} \quad C'_3 + C_5 = C_{a4} + C_{b5} + C_{a6} \quad C_4 = C_{b4} + C_{a5}$$

Inverter 6 is a load to the latch at the destination that captures the Boolean-level. The latch must be loaded proportionally to the same ratio as the first latch, and here scaling the inverters of the latches becomes the issue. It is convenient to introduce a scale factor f_i, where $i = 0$ through 6, that specifies the size of inverter i, in the unit of the minimum-size inverter. The minimum-size inverter consists, for instance, of the minimum-size NFET and the second minimum-size PFET that the CMOS processing technology allows. This is not a precise definition, but the arbitrariness does not matter. Inverter i consists of f_i NFETs connected in parallel and the same number of PFETs connected in parallel. The current drive capability of the size f_i inverter is f_i times that of the minimum-size inverter. The capacitances C_{ai} and C_{bi} are also f_i times those of the minimum-size inverter. Inverter 2 of the latch that sources the signal to the tested inverter has the size ratio $C_{a3}/C_{a2} = f_3/f_2$ to the tested inverter 3. Then the loading condition must be reflected to the loading of the destination latch. We must choose inverter 6 size by $C_{a6}/C_{a5} = f_6/f_5 = f_3/f_2$. The similar loading condition must be matched at node N_3. The size ratio f_4/f_3 must reflect the condition of driving node N_0 by the signal source. The signal-source inverter 0 has the scale factor f_0, and that must be chosen such that $f_1/f_0 = f_4/f_3$. If the signal-source inverter provides effectively a voltage-drive condition at node N_0, $f_1 \ll f_0$ and then $f_4 \ll f_3$ by the requirement: inverter 4 is then negligibly smaller than inverter 3. The last condition required is the ratio $f_2/f_1 = f_5/f_4$. The three conditions are required to make the input and the output clocks of the two latches synchronized.

The circuits of Figs.2.15.1(c) and (d) are used to determine the inverter delay. The delay measurement can be generalized to any gate or any digital circuit substituting inverter 3. This circuit includes the latches that work as clocks at the input and at the output, but delay measurement does not require node N_2 and node N_3 voltage-waveform determination. The intelligent capability of observing the node-voltage waveform and to determine delay time of the Boolean-level wave is carried out by the latches. The measurement scheme becomes especially simple if the drive capability of the signal source is so high that it may be considered as a voltage source. Then inverters 4, 5, and 6 are negligibly small compared with inverter 3 (or generally the output stage of the measured gate chain). Then there is no ambiguity in the specification of delay measurement. If a gate other than an inverter or a gate chain is to be measured, it must be enabled. If the gate circuit makes a converging tree to the output, only one destination latch is required, but many input latches, all clocked by the same clock signal, are required. If the gate circuit makes a diverging tree from the input to the output, many output latches are required, but only one input latch is required. Generally, multiple input and multiple output latches are required. At the interface of the latches (both at the input and the output) to the tested circuit, the drive capability and capacitive loading conditions must be matched. The tested circuit or the gate may contain parasitic capacitances of wire and so on. If the extra capacitance is included, the scaling requirements must be changed accordingly.

The delay time of the inverter is determined as follows. First, we need to determine the time required for the digital signal to go through the source latch. In the circuit of Fig.2.15.1(b) the switch is set at A (V_{DD}) for a long time. The inverter chain is set at the initial condition, in which nodes N_2 and N_4 are at the HIGH Boolean-level and nodes N_1, N_3, and N_5 are at the LOW Boolean-level. Inverter 0 drives node N_0 to the Boolean LOW level. If the switch changes to position B, the input of inverter 1 goes down, and the signal begins to propagate to the nodes N_1, N_2, ..., successively. At time t_1 the switch is turned to position C. Then the inverter chain is sourced by the latch, created by closing a loop including inverters 1 and 2. The latch settles either of the two states: if the time after switch A $\to$ B transition to B $\to$ C transition is long, node N_0 is LOW and N_1 is HIGH, and vice versa. The measurement is of a trial and error type: many measurements are carried out, and the boundary between the two regimes is determined. At $t_1 = T_1$, a metastable state of a latch is reached.

Let us keep the switch at B position at all the later times. The digital signal front propagates through the inverter chain to the downstream. While keeping the input condition, the switch at the input of inverter 4 that was originally at position D is switched to position E at time t_0. If t_0 is short, node N_4 will settle at the HIGH level, and N_5 will settle at the LOW level. If t_0 is long, node N_4 settles at the LOW level, and node N_5 settles at the HIGH level. There is a boundary between the two regimes at $t_0 = T_0$. At $t_0 = T_0$, the latch goes into a metastable state. The delay time T_D of the tested inverter (or the gate chain) is then defined by

$$T_D = T_0 - T_1$$

Figures 2.15.2(a)-(c) shows the node-voltage waveforms of the circuit of Fig.2.15.1(b). The closed circles show node N_2 voltage, and the open circles node N_1 voltage. In Fig.2.15.2(a) the switch changes position from B to C at $t_1 = 3.5$. Since charge sharing among capacitors C_0 and C_2 occurs instantly through the ideal switch, V_2 changes stepwise from $V_2(t_1-0)$ to

$$V_2(t_1+0) = \qquad (2.15.1)$$
$$= [C_2 V_2(t_1-0) + C_0 V_0(t_1-0)]/(C_2+C_0)$$

$V_2(t_1+0)$ and $V_1(t_1+0)$ are the measures of input information delivered to the latch by $t = t_1$. If more information is delivered, the latch state is more likely to change. If $t_1 = 3.5$, the information is not enough, but if $t_1 = 3.7$, it is enough to change the state. The boundary between the two is about $t_1 = T_1 = 3.6$.

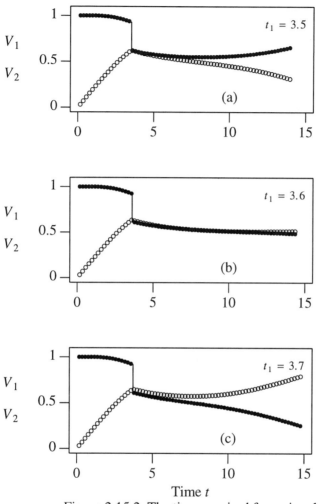

Figure 2.15.2 The time required for a signal to go through a latch

The node voltage V_2 steps of Figs.2.15.2(a) to (c) are the signature of transfer of control of the latch circuit from the signal source to the latch itself. Since node N_2 is farthest away from the source, less new information arrived there, but the node holds more past information. The voltage averaging by Eq.(2.15.1) and subsequent quantization are the latch's intelligent function. The following points are important. The step generation by the charge transfer among the node capacitors adds to the clear definition of the time when the input signal reaches the gate. The creation of a voltage step shows that the quantum-mechanical delay measurement has a rational connection to the classical delay definition in the limit as the maximum step height (from the Boolean LOW to the Boolean HIGH voltages). Indeed, the classical definition has no ambiguity if the transition waveforms are crisp. Isolation of the latch circuit and the voltage step generation means that the two nodes of the latch circuit became physically equivalent. Before the switch position change, the time references of the three nodes were different. After the switch position change, the *fastest* and the *slowest* nodes merge to make a single node, and the latch has the *middle* node and the *merged* node. The time references of the two nodes became the same by the averaging of the node voltages by the charge sharing. The signal and its complement become available as a pair of synchronized stepfunctions. This point is discussed in the next section.

In Sections 1.08 to 1.11 I summarize the problems of the classical delay-time definition. In this section, I mention that the problems were resolved by introducing the quantum-mechanical delay-time measurement. Let us observe how the problems were resolved:
(1) Uncertainty in the delay time due to the unspecified logic threshold voltage.
This difficulty was removed by the introduction of the quantum-mechanical delay measurement. The delay time is measured from a latch to a latch and not from the input to the output of a gate. A cost of this new definition is that a delay time cannot be assigned to each gate, and that creates difficulties in compiling a design data. This issue can be resolved, however, by adding qualifications to the classical gate delay table, by noting that the sum of the classical delay times of the gates in a logic chain is only an approximation, often quite a poor one, and by noting that the tabulated delay data should be used with caution. The problem already existed in low-speed circuits but has not been recognized.
(2) Nonmonotonous node-voltage waveforms.
The destination latch makes a decision about whether the signal arrived. Then, whatever waveform it may receive, the latch makes a decision, except for the rare case of metastability. The waveform issue goes to the background by latch's *intelligence* or, more precisely, the latch's autonomy in making decision. This means that the latch may make a wrong decision, similar to many intelligent decision makers, and we use the circuit with caution. The source of this difficulty is not in the circuit level but rather in the higher, architecture level. Indeed, the nonzero probability of making the wrong decision is the probabilistic nature of the individual quantum mechanical measurement, but that is consistent with the basic physics that governs a digital circuit.
(3) Backward motion of the information-carrying object.
This problem was resolved by identifying that it is a close analog to the quantum-mechanical tunnel effect. This phenomenon provides an easy model for quantum-mechanical tunneling. The backward motion does not result in a breakdown of causality in the latch-based delay-time

measurement. The destination latch gets the signal always later than the source latch, and the positive delay time is guaranteed.
(4) The nonexistence of a proper delay-time definition in dynamic gates. If a signal is introduced by opening a latch by a clock, and the delay time is measured by the second latch, the signal is a version of the clock, enabled or disabled by the other signal. This idea is consistent and is more natural if it is considered on the quantum-mechanical basis.

Although the problems of delay time were all resolved, is the new definition more convenient and realistic than the classical one? The quantum definition is certainly more realistic in the digital circuit than the classical delay-time definition that contains built-in inconsistencies. Yet the definition is less convenient than the classical definition. I point out that the classical definition was really a groundless, arbitrary definition, and that becomes especially clear in high-speed circuits. In a way, we lost something we really never had before, and instead we gained a valid perspective. To resolve the convenience problem, the sort of CAD that must be used in the design becomes the issue. This problem is discussed in Section 3.15.

2.16 Design Guidelines for Ultrafast Circuits

To design ultrahigh-speed circuits, practically all design data and every design tool, such as FET characteristics and the circuit simulator, are uncertain. To design a circuit subject to these conditions, we need certain qualitative guidelines to choose the component connectivity and the device characteristic. Many, indeed most, of these additional requirements originate from the quantum-mechanical nature of a digital circuit delay. These problems are summarized and discussed in this section. To design the highest-possible digital signal-processing speed, we must address the following issues from the engineering viewpoint:

(1) Use simple circuits, and minimize the number of inputs to a logic gate, so that the gate's response time becomes the minimum by its structure and not by its design parameter (like FET size). Simplicity of circuit means, especially, to avoid stacking FETs in a series. This is important to prevent the back-bias effect of the FETs during switching, whose sources are not directly connected to the power or the ground bus (Brews, 1981). In a deep submicron CMOS technology the back-bias effect is quite significant because the tubs are highly doped to prevent the short-channel effect (drain voltage dependence of the FET conduction-threshold voltage V_{TH}). Minimization of the back-bias effect and minimization of the short-channel effect conflict with each other. Using this guideline, the fastest circuit can be built by using only inverters and transmission gates that are driven by a voltage source (a large inverter buffer).

(2) Resistive loading or conflict of drive of a node does reduce the signal amplitude but does not reduce the speed of the circuit

significantly. Resistive loading and the drive conflict, however, increase DC power. If that is justifiable, this is a good way to increase speed. An example of the design guidelines (1) and (2) is the circuit structure of a high-speed latch. The conventional CMOS latch built using two tristatable inverters and an inverter shown in Fig.2.16.1(a) is slow because of the stacked FETs of the tristatable inverters. The circuit of Fig.2.16.1(b) is faster than that if the signal is supplied from a large inverter buffer. The circuit of Fig.2.16.1(c) is the fastest because it is simple, although the direct output is logically inverted. An inverting output creates no problem if the latch is used in the edge-triggered master-slave configuration or if the downstream logic is properly designed. Inverter B is much smaller than inverter A and the transmission gate: its role is to maintain the input node of the inverter A at the stable Boolean-level.

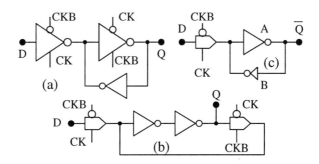

Figure 2.16.1 Various latch structures for hogh-speed data transfer

(3) Avoid generating a do-undo pair process and the associated nonmonotonic node voltage waveform. Such an waveform makes the signal processing delay long and uncertain. A do-undo process occurs when a single gate receives two input-level changes that maintain the output level the same after all the transients are over. If the two changes occur in the timing order to maintain the gate kept disabled, no extra waveform feature is generated at the output, but if the timing order enables the gate temporarily, an isolated pulse is generated. The width of the isolated pulse depends on the timing between the two input pulses. This process is called a *hazard*. A hazard can be removed by adding complexity in the logic circuit, but that is not desirable for the speed (McCluskey, 1986). The only way to remove hazard in the high-speed circuit is to adjust the circuit delay such that the gate input timing is correct.

(4) If a nonmonotonic node waveform cannot be eliminated, it is necessary to convert it to a monotonic waveform within the minimum number of propagation stages after its generation. If the extra pulse

cannot be removed, the timing of the two input signals is adjusted to make the isolated pulse width to the minimum and to design the gate that receives the signal to erase the extraneous feature. In Fig.2.16.2 the NAND2 gate generates a hazard. To deal with this problem, decrease the arrival-time difference ΔT to the minimum, and make NFET MNA small, so that the narrow pulse is stopped by the inverter. Because this is an important issue, we provide further discussion.

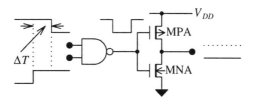

Figure 2.16.2 Static hazard and its erasure

A hazard is an interference in a digital circuit. The interference is demonstrated by a circuit in which a single signal is split into two (or more, in general), and the split signals are propagated along the separate paths. The lengths of the paths that determine the relative delays of the split signals matter critically. At the downstream of the paths, the signals are combined by a multiinput gate. Both constructive and destructive interference can be produced. This type of structure is called a *recombinant fanout*. There is an outstanding feature in digital circuit interference. In digital interference, two analog voltages that may not have definite Boolean-levels are combined, and then the sum is quantized. At the input of a multiinput gate, the input signals have analog voltage values. After they are added by the gate circuit and amplified, a quantized Boolean-level emerges at the output of the gate. The input analog signal is equivalent to a probabilistic superposition of the HIGH and the LOW Boolean-level voltages.

The best-known circuit structure that generates a hazard is a circuit creating the destructive interference shown in Fig.2.16.3. In this circuit a stepfunction signal is split into two paths having different delays. In a steady state the two inputs of the NAND gate have complementary Boolean-levels, and therefore the steady-state output of the gate is HIGH. Since the second branch has the extra inverter chain delay, the following process takes place. Suppose that V_{N1} is at the LOW logic level. Then V_{N2} is in the HIGH logic level. Suppose that V_{N1} makes a LOW to HIGH transition. There is some time before the node N1 upgoing signal reaches node N2. During that time both inputs of the NAND2 gate are HIGH, and the output of the gate becomes LOW. The NAND2 gate generates a return to one pulse at the output, and the width of the pulse equals the delay of the cascaded inverters between nodes N1 and N2.

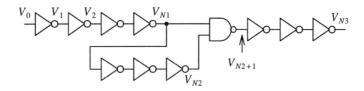

Figure 2.16.3 The generation of an extra transition edge

Figures 2.16.4(a)-(c) shows the results of a numerical analysis. The number of inverter stages between node N1 and node N2 is changed from 5, 3, and 1, in the order of Figs.2.16.4(a), (b) and (c).

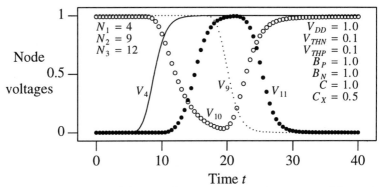

Figure 2.16.4(a) The generation of extra transition edge at the recombinant fanout

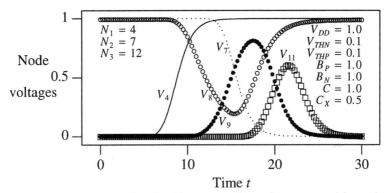

Figure 2.16.4(b) The generation of extra transition edge at the recombinant fanout

Quantum-Mechanical Gate Field

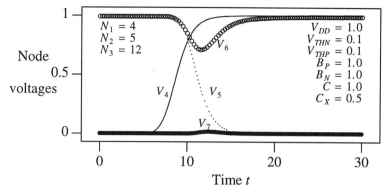

Figure 2.16.4(c) The generation of extra transition edge at the recombinant fanout

The NAND2 gate is a conventional CMOS NAND2 gate, having two parallel-connected PFETs to pullup and two series-connected NFETs to pulldown. The inverters and the NAND gate are built from the same NFET and PFET. The gate's internal node between the two series-connected NFETs is not a digital node, to which a smaller capacitance than C, C_X, was assigned. The two NAND gate-input-node waveforms are shown by the solid and the dotted curves, and the NAND2 gate-output-node waveform is shown by the open circles. Whether or not the NAND2 gate output generates a pulse having CMOS logic levels after one stage of buffering is the criterion. The buffered signal is shown by the closed circles.

Figure 2.16.4(a) shows that a healthy CMOS level isolated pulse is generated if the number of the inverters between nodes N1 and N2 is 5. Figure 2.16.4(b) shows the result for a circuit having three stages of inverters between the two input nodes of the NAND gate. We observe a marginal generation of an isolated pulse at the one-stage buffered node 9 (closed circles). The RZ pulse keeps decaying on further buffering, as is observed from the waveform of node 11 (squares). Figure 2.16.4(c) shows the results for a circuit having only one stage delay inverter. The output-node voltage of the NAND gate decreases temporarily, but it is not sufficient to generate any isolated pulse at the buffered node 7 (closed circles). The delay time of five or more inverter stages is required to generate a propagating isolated pulse.

(5) A design verification must be carried out by using the circuit-level simulator. The gate-level simulator at present does not have high enough accuracy to detect the circuit problems. A hazard pulse in a reasonably well-designed circuit never reaches the full-logic swing, and therefore a logic simulator is not adequate. Node-voltage waveforms must be determined accurately.

(6) Optimize the circuit performance as a whole rather than each of

its building blocks. The required processing time can be squeezed out if the entire circuit is designed at once. This method has the following details. In a higher-speed circuit the node voltages observed over time are less frequently in the ranges that have the definite Boolean values. This is acceptable over a several stages of gates in a well-designed circuit but not in a complex circuit over indefinitely many stages. We need to reestablish healthy stepfunction waveforms. The only way to do so is to insert a static latch having positive feedback in the signal path. A static latch has effectively infinite gain and wace-reshaping capability. The latch interprets the nonideal node waveform and extracts the Boolean-level information as a stepfunction. The schematic diagram of a logic circuit is as shown in Fig.2.16.5.

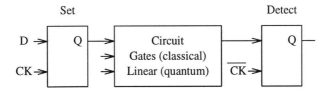

Figure 2.16.5 A new way to model a sequential-logic circuit

If these viewpoints are rephrased from our theory, several conclusions are reached:

(1) The maximum circuit speed is reached by minimizing the latch to latch delay time.

(2) The circuit between the two latches need not be constructed from logic gates, which are often considered as the minimum unit of logic function. Between the two latches the digital signal need not be represented by a classical particle, but it can be a quantum-mechanical wave function, and therefore there is no need to use logic gates to reproduce the classical particle nature of the digital signal at every stage. This conclusion has wide ramifications for digital-circuit speedup:

(a) The most important component of a digital circuit is the latches that execute digital quantization, thereby extracting the classical particle nature of the digital signal. The latch must have high gain and must be fast, but not too fast to be sensitive to capture the spurious features of the signal. The latch must have the capability of *interpreting* the input waveform.

(b) The logic HIGH and LOW levels of the nodes of the circuit between the two latches need not have a definite Boolean-level for the correct operation of the circuit. Any circuit that is able to generate a properly quantizable output waveform at the destination latch is acceptable. Figure 2.16.6 shows logic amplitude schematically.

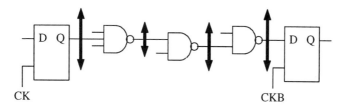

Figure 2.16.6 A logic amplitude of the tunnel-effect logic

(c) If an independent stage of gate or amplifier is used, the delay cannot be less than the certain minimum. To avoid this problem, the circuit of (b) can even be a passive circuit. If signal amplitude is not the issue, the pole-zero compensated voltage divider has circuit-theoretically zero delay time. For the speedup scheme to work effectively the circuit must satisfy several requirements. If it is a single-stage circuit, it must start responding to the input change instantly: the output voltage immediately following a stepfunction input voltage is proportional to Δt, where Δt is the time referred to the input stepfunction edge. Delayed response, like being proportional to $\Delta t - t_A$ or $(\Delta t)^2$, . . . , is not desirable. The one-stage circuit modeled as an amplifier must be not in class C but in either class A, AB, or at least B. Since a class A amplifier consumes standby power, a class AB or class B circuit will be the best choice. This design strategy is, in effect, to tunnel the information-carrying *pseudoparticle* from the set latch to the detect latch of Fig.2.16.5. The design strategy has its origin in the quantum-mechanical tunnel effect, which takes place instantly. The quantum-mechanical tunnel effect occurs instantly, and therefore the circuit-design strategy is to simulate the structure by a circuit. In the tunnel effect, the observed result is created by the intervention of the test equipment. This strategy may be rephrased as catching the early warning and constructing the result from it. The condition necessary for creating the analogy of the tunnel effect in an electronic circuit is that the impulse response is proportional to Δt. This is an important new viewpoint for looking at the ultrafast digital circuit. The requirement of the fastest digital circuit is consistent with the fundamental property of an electron triode amplification mechanism that leads to the gain-bandwidth product (Shoji, 1996). Because of the limit set by the product, an infinitely fast transient is not allowed in any electronic circuit. The circuit response has the essential gain-bandwidth limit, but in a real quantum-mechanical world the effective gain-bandwidth product is infinity. Realistic voltage stepfunction has small but nonzero rise time $t_{R/F}$, and the stepfunction response of a digital gate is a function of $\Delta t - (V_{TH}/V_{DD}) t_{R/F}$ rather than proportional to Δt (Section 2.05). To get a significant circuit speedup, the term $(V_{TH}/V_{DD})t_{R/F}$ must be nulled. One way (but not necessarily the only way) is to use the latch circuit shown in Fig.2.16.7 to drive the downstream gate starting at either V_{THN} or $V_{DD} - V_{THP}$: the circuit responds instantly to the stepfunction input change. Another way is to fabricate both types of FETs on the respective tubs that can be biased independently and, by applying proper polarity bias, to reduce the FET conduction-threshold voltage using the back-bias effect.

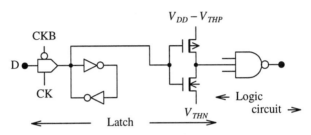

Figure 2.16.7 The rapid start of static logic

The logic-speed gain attainable by the pseudoparticle tunneling logic is estimated as follows. Let the logic circuit between the set and the detect latches of Fig.2.16.5 be a pullup and pulldown symmetrical inverter, whose FETs are modeled by the collapsable current generator model. The classical inverter delay time (0 to 100%) to the stepfunction input is given by $T_{DC} = CV_{DD}/[G_m(V_{DD} - V_{TH})]$, where C is the node capacitance, G_m is the transconductance, and V_{TH} is the FET conduction-threshold voltage. The delay time of the linear amplifier T_{DQ} is defined by driving the input by a stepfunction having amplitude ΔV_1 ($< V_{DD}/2$) and determining the time when the output voltage falls by ΔV_1. Since the transconductances of both the PFET and the NFET add together, the current is $2G_m \Delta V_1$ and the charge to be drained is $C \Delta V_1$. Then the delay time is $T_{DQ} = C/(2G_m)$. The ratio of the two is

$$T_{DQ}/T_{DC} = (1/2)[1 - (V_{TH}/V_{DD})] \approx 0.5$$

if $V_{TH} \ll V_{DD}$. The speed gain by a factor of 2 is quite significant in a speed-critical ultra large scale integrated (ULSI) circuit, especially the critical paths of a microprocessor.

The impulse response cannot be made proportional to Δt in a multistage logic circuit. Since at least two cascaded logic-circuit stages are required for design flexibility, the best we can hope for is that the response is proportional to $(\Delta t)^2$. This is still quite a significant logic circuit-speed improvement. From the analysis of Section 2.05 the response time of a single-stage inverter is proportional to $[\Delta t - (V_{TH}/V_{DD})t_1]^3$. This is not the condition favorable for creating the information-carrying pseudoparticle tunneling from one latch to the other. The circuit between the two latches must be a linear circuit of at least class B, and this is the crucial difference from the conventional circuit-design strategy. To make a multistage circuit work like a single-stage circuit, nonconventional circuit-design strategy is required. The conventional way is to choose the layout style that minimizes the node capacitance. An ultrafast circuit must be laid out with full understanding of its operation. A few other strategies are to use a pole-zero compensation scheme or use negative capacitance to compensate for the positive-node capacitance. This technique has been considered not practical, but it may be the only solution for speeding up certain critical paths in an ULSI circuit like a microprocessor.

(d) The design of a ultrafast circuit must be carried out including the latches. The delay time within the latch and the switching threshold of the latch must be optimized and held at the optimum level to be consistent with the output response of the combinational circuit, for all the possible process, temperature, and the power-supply voltage conditions. If the circuit is designed that way, the latch-to-latch circuit is the fastest circuit one is able to build and operate. To adjust the threshold voltage of the latch is effectively to choose two different time references of signal introduction at input and its detection at output.

(e) The latches themselves can do many operations, including the shift of the time reference referred to above. Latches can be designed to carry out logic operations of their own, like multiplexing, with no overhead. This circuit structure is different from the cascode voltage-switch circuit (Heller, Griffin, Davis and Thoma, 1984), because the major part of the tunnel effect logic is still carried by the circuit between the two independent latches.

The design strategies summarized before are related to the issues of the circuit level. Above the circuit level, namely, in the system's architecture level the following guidelines apply:

(1) Separate the entire circuit into two or more blocks, such that the high-speed circuit is minimized. By increasing the width of the data by serial-parallel conversion, the complex operations are carried out at the low clock rate. This strategy often creates a peculiar critical path, and therefore the design verification must be carried out exhaustively.

(2) In some circuits, delay of the output signal by one clock cycle, or often in any integral number of clock cycles, creates no problem in the system-level application. Then by delaying the signal it is possible to gain processing time or even margin. If this scheme is used, it is important to maintain the master-slave data-transfer format, even if the processing-delay time of the gate chain exceeds the clock period. Eliminating either the master or the slave latch is effectively equivalent to using the gate-chain delay time as the memory, and then the delay time

must have the lower bound, as well as the obvious upper bound. To maintain the lower bound in the case of a fast process condition, low operating temperature, and high power-supply voltage, extra design care is required. To set two delay limits is usually not an easy task. If the circuit receives a fixed-frequency external clock, it is possible to create a fixed time-delay time by scaling the clock period by a fixed factor. As I discussed in Section 1.08, the delayed digital signal must have a crisp switching waveform, so that the classical delay time is reasonably well defined.

(3) Dynamic NOR logic gate has a speed advantage above all, a simple structure to eliminate many problems. The output waveform consists of pullup by precharge and pulldown by a bank of NFETs, and the waveform is simple, too. The fastest dynamic logic circuit, called *NORA dynamic logic*, is schematically shown in Fig.2.16.8 (Goncalves and de Man, 1983).

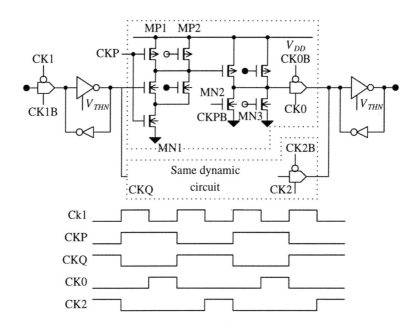

Figure 2.16.8 A method for hiding the precharge time slot of a dynamic gate

The first stage is the fast NFET-based dynamic-logic circuit that has the ground switch. The output of the first stage drives the PFET-based dynamic-logic circuit, and so on, and the types of the dynamic gates alternate. A dynamic logic circuit requires a precharge and predischarge period. The precharge period can be made invisible by providing two dynamic circuits, and they precharge and discharge alternatively, as schematically shown in Fig.2.16.8. A two-stage cascaded NORA logic circuit is driven by the clocks shown in the lower part of the figure. MP1 is the precharge PFET, MN1 is the ground switch of the first stage, and MP2 is to compensate for the leakage current of the circuit, which is not negligible since the input signal begins to drive from V_{THN} and not from zero. MN2 is the precharge NFET, and MN3 is for leakage compensation.

The scheme becomes realistic in the scaled-down CMOS. Using many FETs to speed up a speed-critical paths is now a rational choice. This is an example of an interesting and productive heuristic method for designing an ultrafast digital circuit seeking an analogy to a quantum-mechanical experiment. We consider how an information-carrying elementary particle behaves in the quantum-mechanical experimental setup and try to create its equivalent-circuit model. The circuit of Fig.2.16.8 is closely similar to the double-slit interference experiment of an electron wave. NORA dynamic logic circuits have been avoided by conservative designers as unreliable. This bias must be shed, and the circuit technique must be established now, since the ultimately high-speed CMOS system requires NORA to carry out the complex logic operation.

(4) Some system-level measures for suppressing the IC process condition variations, as well as the IC operating temperature and the power supply voltage variations, are required. It usually is easier to design the analog circuits that carry out the stabilization than to obtain the last few percentages of improvement in digital-circuit speed by other methods.

(5) Using a class AB or class B amplifier circuit to simulate the tunnel effect between the latches has one issue that the designer must be aware of at the time of design verification. This is the initial condition of the circuit before switching. Since the circuit does not start from the definite voltage level as enhancement mode CMOS (that is, class C), the residual effect of the previous switching event must be carefully evaluated. The circuit must be tested for various data patterns for error-free operation.

(6) The power-supply voltage of a CMOS technology is kept at 5 volts for about fifteen years. Now the voltage is being reduced, and associated with it, the following new circuit speedup mechanism emerged. Reduction of V_{DD} allows thin gate oxide, below 100 angstroms. The larger gate capacitance (per unit area) creates larger conductance of the FETs. The node capacitance is a sum of the gate and the interconnect capacitances. Then the fraction of capacitance effective in creating device conductance increases by decreasing V_{DD}. This mechanism makes a CMOS circuit having comparatively more wiring faster. The availability of many levels of metals allows minimization of the series resistance of the interconnect, which has been the significant delay

component in the old generation CMOS (Shoji, 1987).

(7) Distribution of the primary clock must be carried out by using the H-clock distribution scheme, or its electrical equivalent (Shoji, 1992). This is practically the only way to ensure absolute synchronization over the entire chip. Yet the clockedge at the clocked point and at the chip's clock input are different by the buffer and distribution delay. The difference must be set equal for all the interacting chips by using an analog-circuit technique. Some means of deskewing the clock and its complement must be used to generate complementary pair of clocks. The precision delay matching of Fig.2.16.9(a) (Shoji, 1987) or the positive feedback arrangement of Fig.2.16.9(b) becomes necessary to deskew the clock pair. The circuit of Fig.2.16.9(b) uses the latch circuit, which tries to pull the time reference of the two nodes close together.

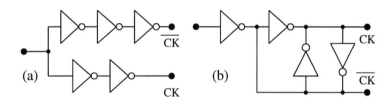

Figure 2.16.9 Deskewing clocks

(8) The clockedge that controls the latch must be adjusted to balance the processing load between the latches. In a synchronized system the maximum speed is attained only by providing several clocks that are centered at the default clockedge but shifted slightly by the buffers that track the process/temperature/supply voltage variation to match the need of the individual clocked point.

Which of the quantum-mechanical (Section 2.09) or semi-quantum-mechanical Boolean-level determination methods (Section 2.15) is used depends on the system. A heavily capacitively loaded node like a memory bitline is measured by the quantum-mechanical method, as is done in DRAM/SRAM readout circuit. The circuit operation is often called *refresh*. The method will be used more and more in high-speed dynamic logic circuits along with Domino CMOS and NORA circuits. Yet the conventional semi-quantum-mechanical delay-measurement method will be popular at least at present. This is because the complexity of the system demands flexibility in circuit design.

2.17 Natural Decay of Composite Excitation

Let us consider an isolated (or RZ) pulse launched to an infinitely long inverter field, shown in Fig.2.17.1. Does this excitation retain its node waveform unchanged forever, as it propagates to the right? Let us consider the simplest case that the inverters are all identical and pullup-pulldown symmetrical. If the height of the input pulse is less than $(1/2)V_{DD}$, it will not even propagate beyond the first stage. If the width of the input pulse is about the inverter delay time or less, it will not propagate more than a few stages. We consider the nontrivial case that the input waveform has full CMOS logic swing from V_{SS} to V_{DD} and the width is at least several times the delay time, as schematically shown in Fig.2.17.1.

In Fig.2.17.1 the input voltage V_0 was zero before $t = 0$, and it makes a stepwise LOW to HIGH transition at that time. The input upgoing transition edge should move at the velocity determined by the gate field parameter only after several steps of propagation, as was demonstrated in Section 1.06. In Section 1.06, we considered the velocity of a single voltage stepfunction. In Fig.2.17.1 the second, input downgoing edge follows some time after the first upgoing edge. The velocity of the first upgoing and the second downgoing edges at the location far away to the downstream must be equal to maintain the same pulse shape forever. This is reasonable because the two wavefronts should be the same fronts, if we observe them at an even-numbered node-index location and at an odd-numbered node-index location, respectively. If this were the case, an isolated pulse having arbitrary width will travel forever, and it must have the same velocity as the stepfunction wavefront. By a casual observation this conclusion appears inevitable.

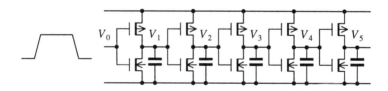

Figure 2.17.1 The decay of an isolated excitation

The velocity of an isolated pulse in an infinitely long gate-field is, mathematically, the inverse of the eigenvalue of the problem, which has the node-voltage profile of the inverter field as the corresponding eigenfunction. The eigenfunction may have great varieties: the length of the flat region that retains the same top (or bottom) phase should not affect the velocity, or the eigenvalue. This is mathematically very strange: then the velocity, or the eigenvalue, must be infinitely *degenerated*. There must be something quite wrong in the previous reasoning. Are the velocities of the front and the rear edges of a RZ pulse really equal? If the distance between the two fronts is long, there is no doubt that the velocities are the same. This statement does not ensure that the velocity is the same if the distance is not infinitely long. Whether the region is long or short depends on a subjective judgment. There is a problem hidden here. The best way to examine whether the front and the rear velocities are equal is to simulate the long chain of inverters.

Figure 2.17.2 shows the node-voltage waveforms of an inverter field that is 400 stages long. The gradual-channel, low-field FET model is used, and the parameter values are given in the figure. Since the input node 0 is an even-numbered node, the waveforms at the later even numbered nodes are shown in the figure. The gate-field is driven by a crisp pulse that has initial width 10. The input upgoing pulse width decreases first gradually and then successively more rapidly, and the isolated pulse decays altogether at the 398th inverter. This observation shows that the trailing edge moves slightly faster than the leading edge, the trailing edge catches up the leading edge gradually, and the pulse width decreases as it propagates. The rate of pulse-width decrease of a moderately wide pulse is so small that the pulse decay is not recognizable by a casual observation of the simulation results.

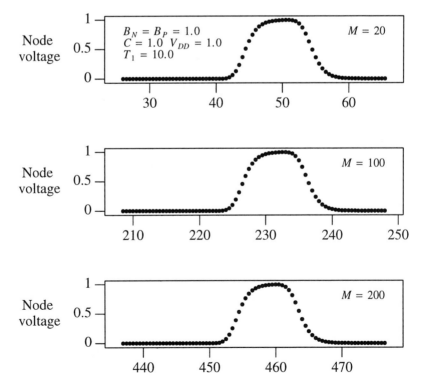

Quantum-Mechanical Gate Field

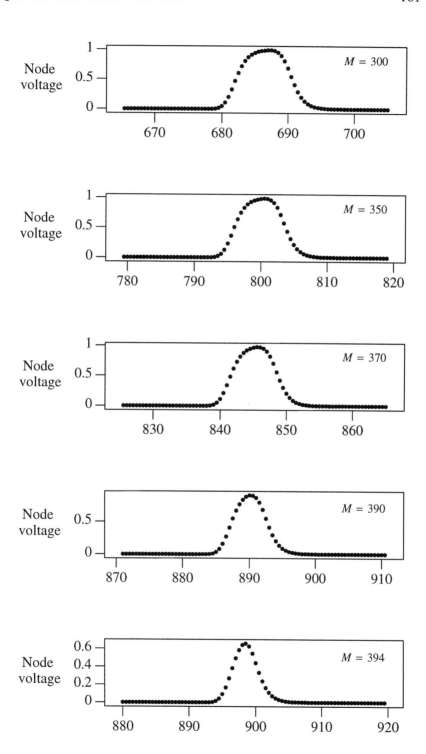

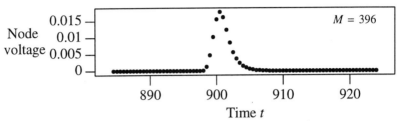

Figure 2.17.2 The decay of a propagating RZ pulse

Figure 2.17.3(a) to (c) shows the decrease of the pulse width versus the node index of the gate-field during the propagation. The pulse width is determined classically, at the level of $V_{DD}/2$.

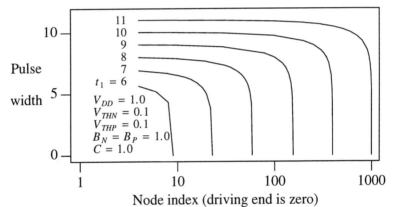

Figure 2.17.3(a) The decrease of a pulse width as it travels an inverter chain

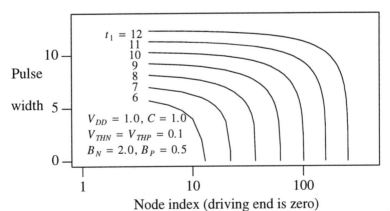

Figure 2.17.3(b) The decrease of a pulse width as it travels an inverter chain

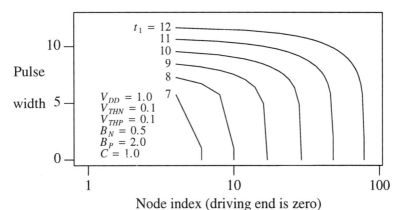

Figure 2.17.3(c) The decrease of a pulse width as it travels an inverter chain

The initial pulse width, t_1, of Fig.2.17.3(a) was changed from 6 to 11. A modestly long pulse having $t_1 = 11$ decays after 1,000 stages of propagation. No practical logic circuit used at present has such a single long propagation path. Figure 2.17.3(a) is for a pullup-pulldown symmetrical inverter field, Fig.2.17.3(b) is for an inverter field having stronger pulldown, and Fig.2.17.3(c) is for an inverter field having stronger pullup. We observe a significant difference in the number of stages required to decay a pulse having the same initial width, and a remarkable deviation from the case of a symmetrical inverter, even if the geometrical averages of B_N and B_P remain the same. A symmetrical inverter supports a pulse for the longest time, subject to the B_N - B_P comparison standard.

The relationship between the number of stages required to collapse a pulse and the initial width of the pulse is plotted in Fig.2.17.4. The relationship is exponential. In a symmetrical inverter the dependence can be approximated by

$$N_{max} \approx \text{const.} \exp[(BV_{DD}/C)t_1] = \text{const.} \exp(t_1)$$

for the example of Fig.2.17.3(a) and in the asymmetrical inverter chain of Fig.2.17.3(b) and (c)

$$N_{max} \approx \text{const.} \exp[(B_{min} V_{DD}/C)t_1] = \text{const.} \exp(t_1/2)$$

where N_{max} is the number of stages required to decay a pulse, and B_{min} is the smaller one chosen from B_N and B_P. We note $B_{min} = 0.5\sqrt{B_N B_P}$ in the examples. The factor 1/2 in the exponent is due to that. If we remember that the last phase of the node-voltage pulldown waveform is approximated by

$$V_0 \to \text{const.} \exp[-B_N(V_{DD} - V_{THN})(t/C)]$$

and since $V_{THN} \ll V_{DD}$, decay of the pulse must be due to the interaction between the two transition edges of the RZ pulse.

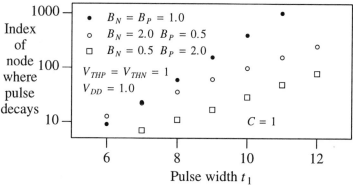

Figure 2.17.4 The number of stages of propagation required to decay a pulse

Figures 2.17.5(a) and (b) show the waveforms of nodes 0, 1, 2, 3, and 4 for an inverter chain that has different pulldown and pullup capabilities. The parameter values for Fig.2.17.5(a) are

$B_N = 2.0 \quad B_P = 0.5 \quad C = 1.0 \quad V_{DD} = 1.0 \quad V_{THN} = V_{THP} = 0.1$

and the initial pulse width was chosen at $t_1 = 6.0$.

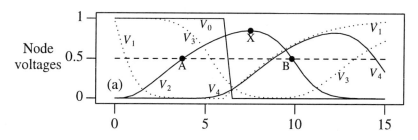

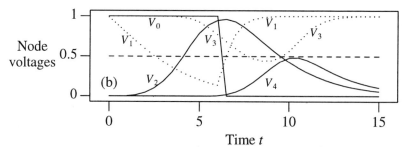

Figure 2.17.5 The mechanism of an RZ pulse decay in an asymmetric inverter chain

Referring to Fig.2.17.5(a), the first inverter pulls down rapidly on input pullup (V_1), but the second inverter pulls up slowly (V_2). At point A, V_2 reaches the logic threshold voltage we assumed, $V_{DD}/2$, and at point X it reaches the maximum and changes from pullup to pulldown. At point X, V_2 has not yet reached V_{DD}. Then the subsequent pulldown process to the logic-threshold voltage $V_{DD}/2$ takes accordingly less time. Pulse-width AB is less than the initial pulse-width, 6. The same process repeats, and the pulse width decreases as it propagates to the right. The crucial issue is that V_2 starts a LOW to HIGH transition from $V_2 = 0$, which has been set by the initial condition of the circuit. The subsequent HIGH to LOW transition caused by the second front launched to the field starts at the initial voltage less than V_{DD}, and that is the reason that the downgoing transition has smaller delay time than the preceding upgoing transition. This explanation highlights the effects of the slow pullup and fast pulldown caused by the different initial conditions, as the dominant cause of the pulse-width reduction. According to this mechanism, the longer time constant of pullup or pulldown dominates in the exponentially dependent effect, since the longer time constant determines the deviation of the transition initial condition more decisively. In addition to this principal mechanism, there is a built-in asymmetry mechanism of the pulse-width alteration because the switching-threshold voltage of the inverter is less than the logic-threshold voltage $V_{DD}/2$. Yet the pulse-width reduction from V_1 to V_2 dominates over the pulse-width increase from V_0 to V_1, and as the cumulative effect from V_0 to V_2, the pulse width decreases.

Figure 2.17.5(b) shows the case where the inverter pulls up rapidly but pulls down slowly. The parameter values are as follows:

$$B_N = 0.5 \quad B_P = 2.0 \quad C = 1.0 \quad V_{DD} = 1.0 \quad V_{THN} = V_{THP} = 0.1$$

The mechanism for reducing of the pulse width is the same. In this figure, the V_1 downgoing pulse is obviously narrower than the V_0 upgoing pulse. The V_2 upgoing pulse, although wider than the V_1 downgoing pulse, does not make up the V_0 to V_1 pulse-width reduction. The pulse decays very rapidly, and the V_4 upgoing pulse does not exceed the inverter switching-threshold voltage V_{GSW} any more. The mechanism of pulse decay is now clearly identified. This effect has a number of interesting consequences, and therefore I present a few more illustrative examples.

In Fig.2.17.6 the decay process of the three isolated pulses in sequence, as they propagate down the gate-field, is shown. M is the index of the nodes, counted from the driving end, and it is an even number. The pulses are a set of reasonably well-isolated pulses, but still their leading edge and the trailing trailing edges overlap, and the pulse-decay mechanism is working. Note that in the waveform plot (voltage versus time), the leftmost pulse (arriving at the earliest time) was launched first, and the rightmost, last. Before the first pulse arrives, the node voltage was exactly zero. When the second pulse arrives at the same location, the node voltage is small but positive. Then the pulse-width reduction mechanism works more effectively to the first pulse than to the second pulse, and the first pulse decays first. As the first pulse decays, the second and third pulses get into the same relation, and the second pulse decays. Thus the multiple pulse decays in the order of the launch.

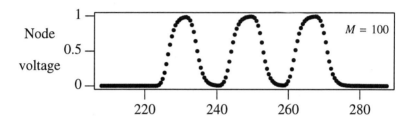

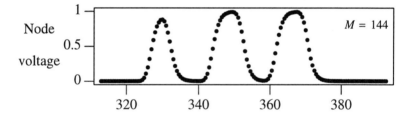

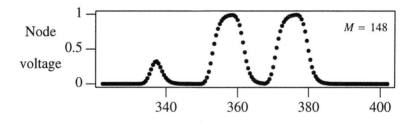

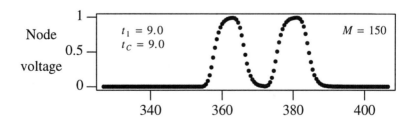

Quantum-Mechanical Gate Field

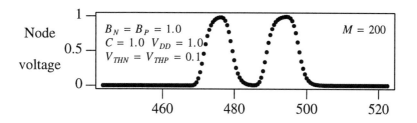

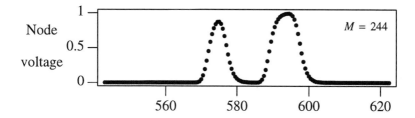

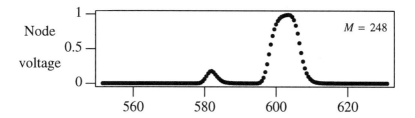

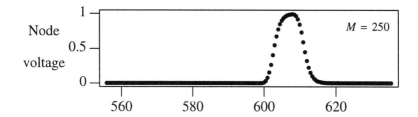

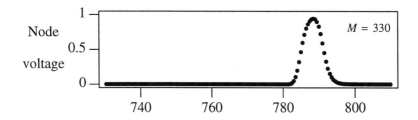

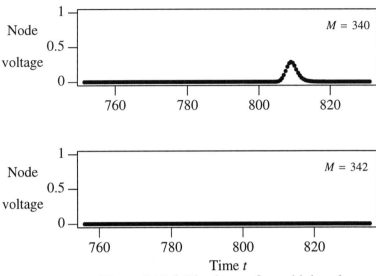

Figure 2.17.6 The decay of a multiple pulse

If they are a set of isolated pulses leading a stepfunction, we see a similar mode of decay, as is shown in Fig.2.17.7. The first-arriving pulse, which is farthest away from the stepfunction pulse, decays first, and then the pulse that immediately leads the stepfunction decays. This mode of decay is clearly analogous to the pulse-decay mode of Fig.2.17.6. Only the stepfunction is stable, and it keeps propagating.

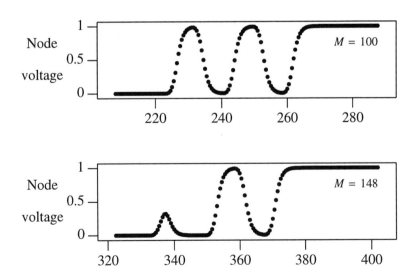

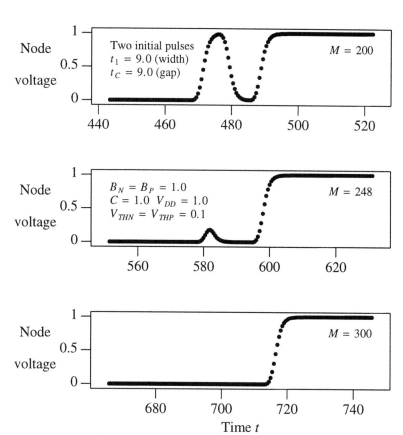

Figure 2.17.7 The decay of a multiple stepfunction

Figure 2.17.8 shows the decay of a combination of two RZ pulses following a stepfunction. In the decay process of multiple pulses, the first-arriving structure, if it is an isolated pulse, always decays first, and the same sequence follows. This is consistent with the previous observation. The mechanism of the pulse trailing-edge and leading-edge interaction was established.

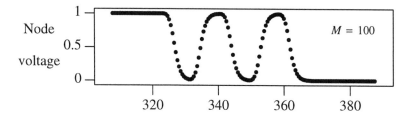

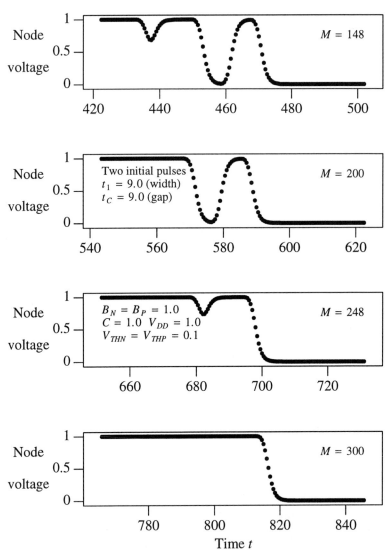

Figure 2.17.8 The decay of a multiple stepfunction

2.18 A Theory of the Decay of Isolated Pulses

The computer experiments of Section 2.17 can be explained by the following physical model. Suppose that an inverter is in a state in which the input is LOW and the output is HIGH for a long time. The steady state has been established for all the nodes of the gate-field. For simplicity, the inverters are assumed pullup-pulldown symmetrical. If the inverter input pulls up, the inverter output pulls down, starting from the accurate HIGH logic level voltage. After an input-pulse width-time later, the inverter input pulls down. At that time, the output node voltage has not yet reached the accurate LOW logic-level voltage. The inverter output begins to pull up from a voltage higher than the logic LOW voltage. Then the output-pullup transition takes a shorter delay time than the first output pulldown transition, which starts from the accurate HIGH level voltage. Then the downgoing pulse at the output of the inverter has a narrower width than the upgoing pulse at the input. This mechanism works for both the return to zero and the return to one input pulse. The pulse width and the pulse height decrease as the pulse propagates the gate-field. After some stages of propagation the pulse becomes so narrow and the height becomes lower than the gate's switching-threshold voltage, and it cannot recreate a pulse at the outout any more. The pulse collapses. This mechanism works even if the inverters are not pullup-pulldown symmetrical. The following closed-form analysis is valid for this general case. The analysis must be able to separate the built-in asymmetry of the pulse-width variation at the even- and odd-numbered index node by the asymmetry: the effects of the pulse-decay mechanism must be cleanly separated from the asymmetry effect.

Let us analyze this problem as follows. The input waveform to the inverter is approximated by the upgoing square pulse having height V_A and width t_1 shown in Fig.2.18.1(a).

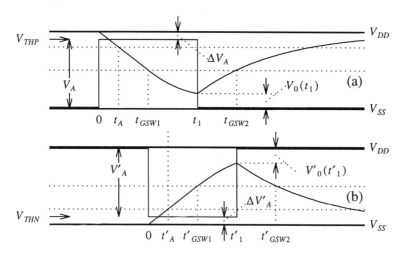

Figure 2.18.1 A guide to the analysis of the decay of an isolated pulse

V_A is close to V_{DD}, and it is in the range $V_{DD} > V_A > V_{DD} - V_{THP}$. It is convenient to define a small parameter

$$\Delta V_A = V_{DD} - V_A \quad (< V_{THP})$$

The following analysis is valid subject to this condition. This analysis practically gives the number of stages of propagation required to collapse a pulse versus the initial pulse width, since if ΔV_A becomes larger than V_{THP} and if realistically $V_{THP} \approx 0.1 V_{DD}$, the pulse decays completely within the following several stages. Origin of time, $t = 0$, is taken at the leading edge of the square-input pulse. At $t = 0$, the output node voltage of the inverter is exactly at V_{DD}. We have the circuit equation

$$C\frac{dV_0}{dt} = -(B_N/2)(V_A - V_{THN})^2 \quad V_0(0) = V_{DD}$$

The PFET is turned off instantly at $t = 0$ since the pulse has zero rise time. This equation is solved as

$$V_0(t) = V_{DD} - (B_N/2C)(V_A - V_{THN})^2 t$$

With reference to Fig.2.18.1, this $V_0(t)$, which is decreasing with time, reaches $V_A - V_{THN}$ at $t = t_A$, given by

$$t_A = \frac{2C}{B_N} \cdot \frac{V_{DD} - V_A + V_{THN}}{(V_A - V_{THN})^2}$$

At $V_0 = V_A - V_{THN}$, the NFET moves from the saturation to the triode region. In the region $t \geq t_A$ and $V_0 \leq V_A - V_{THN}$, the circuit equation is

$$C\frac{dV_0}{dt} = -(B_N/2)[2(V_A - V_{THN}) - V_0]V_0$$

which is solved, subject to the initial condition at $t = t_A$

$$V_0(t_A) = V_A - V_{THN} \text{ as}$$

$$V_0(t) = \frac{2(V_A - V_{THN})}{\exp[[B_N(V_A - V_{THN})/C](t - t_A)] + 1}$$

The switching-threshold voltage of the inverter (of the next stage), V_{GSW}, is determined from the condition that the saturation currents of the NFET and the PFET equal, as

$$(B_N/2)(V_{GSW} - V_{THN})^2 = (B_P/2)(V_{DD} - V_{GSW} - V_{THP})^2$$

We obtain

$$V_{GSW} = \frac{\sqrt{B_P}(V_{DD} - V_{THP}) + \sqrt{B_N} V_{THN}}{\sqrt{B_P} + \sqrt{B_N}}$$

We define the idealized crisp pulse width by setting the logic threshold at V_{GSW}. If $V_0(t)$ equals V_{GSW} at $t = t_{GSW1}$, that is given by

$$t_{GSW1} - t_A = \frac{C}{B_N(V_A - V_{THN})} \log\left[\frac{2(V_A - V_{THN})}{V_{GSW}} - 1\right]$$

At the trailing edge of the upgoing pulse (at $t = t_1$),

$$V_0(t_1) = \frac{2(V_A - V_{THN})}{\exp[[B_N(V_A - V_{THN})/C](t_1 - t_A)] + 1}$$

and the subsequent pullup begins from this voltage level. $V_0(t_1)$ is a small voltage. We assume

$$V_0(t_1) < V_{THN}, V_{THP}$$

and again, if this condition is violated, the pulse vanishes after several more stages of propagation. The pullup process is described, since the NFET is turned off completely, by

$$V_0(t) = V_0(t_1) + \frac{B_P(V_{DD} - V_{THP})^2}{2C}(t - t_1) \quad (t > t_1)$$

$V_0(t)$ increases and reaches V_{THP} at time t_B given by

$$t_B - t_1 = \frac{2C}{B_P} \cdot \frac{V_{THP} - V_0(t_1)}{(V_{DD} - V_{THP})^2}$$

After $t > t_B$, the pullup process is carried out by the PFET in the triode region, and it is described by the equation

$$C\frac{dV_0}{dt} = (B_P/2)[2(V_{DD} - V_{THP}) - (V_{DD} - V_0)](V_{DD} - V_0)$$

The equation is solved subject to the initial condition at $t = t_B$,

$$V_0(t_B) = V_{THP} \text{ as}$$

$$V_0(t) = V_{DD} - \frac{2(V_{DD} - V_{THP})}{\exp[[B_P(V_{DD} - V_{THP})/C](t - t_B)] + 1}$$

If $V_0(t)$ reaches V_{GSW} at $t = t_{GSW2}$,

$$t_{GSW2} - t_B = \frac{C}{B_P(V_{DD} - V_{THP})} \log\left[\frac{2(V_{DD} - V_{THP})}{V_{DD} - V_{GSW}} - 1\right]$$

We define the decrease in the pulse width

$$\Delta t_1 = t_1 - (t_{GSW2} - t_{GSW1})$$

$$= \frac{C}{B_N(V_A - V_{THN})} \log\left[\frac{2(V_A - V_{THN})}{V_{GSW}} - 1\right]$$

$$+ \frac{2C}{B_N} \cdot \frac{V_{DD} - V_A + V_{THN}}{(V_A - V_{THN})^2} + \frac{2C}{B_P} \cdot \frac{V_0(t_1)}{(V_{DD} - V_{THP})^2}$$

$$- \frac{C}{B_P(V_{DD} - V_{THP})} \log\left[\frac{2(V_{DD} - V_{THP})}{V_{DD} - V_{GSW}} - 1\right]$$

$$-\frac{2C}{B_P}\cdot\frac{V_{THP}}{(V_{DD}-V_{THP})^2}$$

In this expression $V_{DD}-V_A = \Delta V_A$ and $V_0(t_1)$ are both small quantities. If we expand the expression into a power series of the small quantities and retain only the first-order terms of the small quantities,

$$\Delta t_1 = T_0 + T_1[\Delta V_A/(V_{DD}-V_{THN})] \quad (2.18.1)$$
$$+ T_2[V_0(t_1)/(V_{DD}-V_{THP})] \quad \text{where}$$

$$T_0 = \quad (2.18.2)$$

$$\frac{C}{B_N(V_{DD}-V_{THN})}\left[\log\left[\frac{2(V_{DD}-V_{THN})}{V_{GSW}}-1\right]+\frac{2V_{THN}}{V_{DD}-V_{THN}}\right]$$

$$-\frac{C}{B_P(V_{DD}-V_{THP})}\left[\log\left[\frac{2(V_{DD}-V_{THP})}{V_{DD}-V_{GSW}}-1\right]+\frac{2V_{THP}}{V_{DD}-V_{THP}}\right]$$

$$T_1 = \frac{C}{B_N(V_{DD}-V_{THN})}[\log\left[\frac{2(V_{DD}-V_{THN})}{V_{GSW}}-1\right]$$
$$-\frac{2(V_{DD}-V_{THN})}{2(V_{DD}-V_{THN})-V_{GSW}}+\frac{2(V_{DD}+V_{THN})}{V_{DD}-V_{THN}}]$$

$$T_2 = \frac{2C}{B_P(V_{DD}-V_{THP})}$$

We continue to determine the output-voltage waveform of the second inverter, which is driven by the first-stage inverter output. We make a simplification shown in Fig.2.18.1(b). The first-stage output waveform is simplified to a square downgoing-pulse waveform that has width $t'_1 = t_1 - \Delta t_1$ and downgoing height $V'_A = V_{DD} - \Delta V'_A$, where $\Delta V'_A = V_0(t_1)$. The same algebraic operations are repeated. The decrease in the pulse width $\Delta t'_1$ is given by

$$\Delta t'_1 = t'_1 - (t'_{GSW2} - t'_{GSW1})$$
$$= \frac{C}{B_P(V'_A - V_{THP})}\log\left[\frac{2(V'_A - V_{THP})}{V_{DD}-V_{GSW}}-1\right]$$
$$+\frac{2C}{B_P}\cdot\frac{V_{DD}-V'_A+V_{THP}}{(V'_A-V_{THP})^2}+\frac{2C}{B_N}\cdot\frac{V'_0(t'_1)}{(V_{DD}-V_{THN})^2}$$
$$-\frac{C}{B_N(V_{DD}-V_{THN})}\log\left[\frac{2(V_{DD}-V_{THN})}{V_{GSW}}-1\right]$$

$$-\frac{2C}{B_N} \cdot \frac{V_{THN}}{(V_{DD}-V_{THN})^2}$$

In this expression $V_{DD} - V'_A = \Delta V'_A$ and $V'_0(t'_1)$ are small quantities. If we expand the expression into a power series of the small quantities and retain only their first-order terms,

$$\Delta t'_1 = -T_0 + T'_1[\Delta V'_A/(V_{DD}-V_{THP})] \quad (2.18.3)$$
$$+ T'_2[V'_0(t'_1)/(V_{DD}-V_{THN})]$$

where T_0 was given before, and

$$T'_1 = \frac{C}{B_P(V_{DD}-V_{THP})}\left[\log\left[\frac{2(V_{DD}-V_{THP})}{V_{DD}-V_{GSW}}-1\right]\right.$$
$$\left. -\frac{2(V_{DD}-V_{THP})}{2(V_{DD}-V_{THP})-(V_{DD}-V_{GSW})} + \frac{2(V_{DD}+V_{THP})}{V_{DD}-V_{THP}}\right]$$

$$T'_2 = \frac{2C}{B_N(V_{DD}-V_{THN})}$$

Let us interpret the physical meanings of the results of Eqs.(2.18.1) and (2.18.3). Their first terms, T_0 for Eq.(2.18.1) and $-T_0$ for Eq.(2.18.3) defined by Eq.(2.18.2), represent the effects of the pullup-pulldown asymmetry of the inverter. This point is clearly observable from the mathematical form of Eq.(2.18.2), since if we set $B_P = B_N$ and $V_{THN} = V_{THP}$, we get $V_{GSW} = (1/2)V_{DD}$, and consequently $T_0 = 0$. If the first terms of Eqs.(2.18.1) and (2.18.3) are added together, the term due to the asymmetry cancels as the cumulative effects of the two-stage propagation, even if the inverters are pullup-pulldown asymmetrical. The rest of the terms depend linearly on the small parameters ΔV_A, $\Delta V'_A$, $V_0(t_1)$, and $V'_0(t'_1)$, and they determine the mechanisms of the pulse-width reduction.

Let us write the rest of the terms of Eqs.(2.18.1) and (2.18.3) as

$$\Delta X = T_1[\Delta V_A/(V_{DD}-V_{THN})] + T_2[V_0(t_1)/(V_{DD}-V_{THP})]$$
$$\Delta Y = T'_1[\Delta V'_A(V_{DD}-V_{THP})] + T'_2[V'_0(t'_1)/(V_{DD}-V_{THN})]$$

The first-stage inverter reduces the pulse width from t_1 to $t'_1 = t_1 - \Delta X$. The second-stage inverter reduces $t'_1 = t_1 - \Delta X$ to $t'_1 - \Delta Y = t_1 - \Delta X - \Delta Y$. We then have

$$t_1 \rightarrow t_1 - \Delta X - \Delta Y \quad (2.18.5)$$

after the two stages of propagation, which restores the pulse polarity. Decrease of the pulse width is $\Delta X + \Delta Y$ per two consecutive stages. Equation (2.18.5) is an underestimate of the pulse-width decrease, however, because the input waveform for each inverter stage was modified to a square pulse having zero rise and fall times. This modification increases the effective FET conductance in the switching, thereby reducing the delay time. Since the real inverter is slower, the decrease of the pulse width will be greater. The real decrease in pulse width will be larger than $\Delta X + \Delta Y$ by a factor f, which is larger than unity but should be close to it. We write

$$t_1 \rightarrow t_1 - f(\Delta X + \Delta Y)$$

and search for the best fit by adjusting f. To compute ΔY, t'_1 is used for pulse width, and $V_0(t_1)$ is used as $\Delta V'_A$. After completion of the two stages of computation, $V'_A(t'_1)$ is used as ΔV_A, $t_1 - \Delta X - \Delta Y$ becomes new t_1, and the process is repeated. Since the computation is complex but straightforward, the iterative computation is carried out by a computer.

Figure 2.18.2 shows the result for a symmetrical inverter field. The number of stages of propagation required to collapse a pulse is plotted versus initial pulse width.

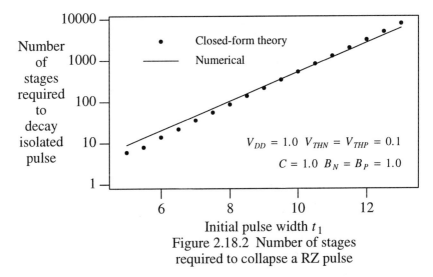

Figure 2.18.2 Number of stages required to collapse a RZ pulse

The initial pulse used in the simulation is a square pulse having width t_1 and height V_{DD}. The rise and fall times are zero. The numerical analysis data and the parameter values were taken from Fig.2.17.4 in the last section. The value of the adjustable factor f was 1.5. Agreement between the closed-form theory and the simulation is quite remarkable, thinking of the simplification used in the theory by approximating a real pulse waveform by a flat-topped square pulse. The dependence is approximately exponential, or the number of stages required to collapse a pulse is proportional to a factor

$$\text{Number of stages} = const \cdot \exp[(B/C)t_1]$$

$$= const \cdot \exp(t_1) \quad (B/C = 1.0)$$

Quantum-Mechanical Gate Field

This simple result is not unexpected from the pulse-narrowing mechanism. Deviation at small initial pulse width in Fig.2.18.2 is due to the strong alteration of a narrow pulse waveform at the early stages. In the numerical analysis, it takes several stages to convert the initial, square pulse to the self-reproducing pulse waveform in the gate-field, and for small t_1 that matters the accuracy.

Figure 2.18.3 shows the results for a pullup-pulldown asymmetrical inverter chain. In this case, a pair of symmetrical parameter choices (A) $B_N = 2.0$ and $B_P = 0.5$ (closed circles) and (B) $B_N = 0.5$ and $B_P = 2.0$ (open circles), were made.

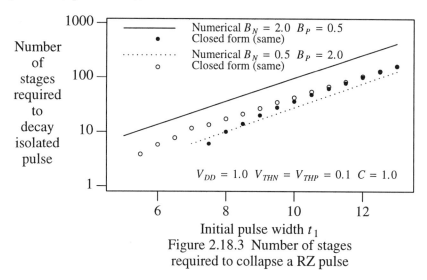

Figure 2.18.3 Number of stages required to collapse a RZ pulse

From the parameter symmetry, the theory should give the same result in the limit as a long propagation. This was indeed the case of the closed-form theory, for $t_1 > 10$. As for numerical simulation, the results are significantly different. This is because the initial adjustment phase of the square pulse to the self-reproducing pulse shape takes several stages of propagation. The theory still gives support to the model in three respects. The number of stages required to collapse a pulse is given by

$$\text{Number of stages} = const \cdot \exp[(\text{Min}(B_N, B_P)/C)t_1]$$

$$= const \cdot \exp(0.5 t_1)$$

Either pullup or pulldown that takes longer time determines the rate of decrease of pulse-width dominantly. The open and closed circles of the closed-form theory are contained in the band between the two curves of the numerical analysis. The slopes of the t_1 dependence, estimated from the two theories, are the same. These three observations correlate the numerical simulation and the closed-form theory well.

The theory of this section is an improvement to the first, very crude theory I presented in an earlier book (Shoji, 1992). This new theory includes gate-fields built from asymmetrical inverters, and it has a clear physical model. Decay of an isolated propagating pulse is now well established. At this point a philosophical proposition is quite tempting. What is permanently stable in the gate-field, a propagating stepfunction, carries infinite *observed* energy with it (Section 1.04). An isolated pulse has only finite observed energy (Section 1.04) and finite lifetime. In elementary particle physics, the truly elementary quarks are permanent existence, but as an isolated object a single quark will carry infinite energy and thus is unobservable. Are these two points just similar but unrelated issues? I suspect they are somehow related: the similarity between the two is too tempting not to encourage some thought about their fundamental existence in nature.

2.19 Mass of Digital Excitation

Mass of a particle is a classical physics concept, but its origin is in the depths of quantum physics. Mass is so fundamental a concept that, as long as we seek for a parallel between a particle as a *mass* point and a digital excitation, we must identify the counterpart of mass of a digital excitation. To do so, the classical approach of constructing a parameter having the dimension of mass from the other gate-field parameters or simply of converting the energy carried by an excitation to mass by the relativity formula is superficial and futile, for two reasons. First, the classical parameters that are simply related to mass, such as force, acceleration, and kinetic energy, are not natural concepts in gate-field, and second, a digital excitation is characterized only by the gate-field parameters but by nothing of its own, and therefore it appears truly an elementary particle. We need to loop for a deep connection between them.

Presently, the theory that successfully explains the origin of mass is the electroweak, theory. The theory predicted the existence of the three intermediate-vector Bosons W^+, W^-, and Z^0 having mass, which are experimentally confirmed, one already known gauge Boson, photon, having no mass and a yet unconfirmed Higgs particle having mass (Aitchison, 1983). The mechanism of the emergence of mass is a phase transition associated with the mechanism called the spontaneous violation of symmetry, or *hidden* symmetry. The mechanism is simple, beautiful, and general. I will show that a simple mechanism that can be considered parallel to that is operative in a gate-field. From this parallelism we learn which excitation in a circuit behaves like a particle having mass and which does not. This is not an arbitrary attempt, since many original ideas of the electroweak, mechanism have their origin in the phase transition in solid-state physics. Some gate-fields indeed go through a phase transition.

We begin with a search for a phase transition in a gate-field. In section 1.12 we showed that a gate-field that is an integration of an

Quantum-Mechanical Gate Field

inverter chain and an RC chain goes through a phase transition at a critical inverter gain (or the resistor loading). The example showed that a gate-field phase transition does exist. Let us consider a simpler gate-field structure than that, shown in Fig.2.19.1. We use the collapsable current generator model of FETs having transconductance G_m and $V_{TH} = 0$. The inverter is pullup-pulldown symmetrical. If the input and the output of a single inverter is connected, the common node settles at $V_{DD}/2$. Each node is connected to a $V_{DD}/2$ voltage source via resistor R. Let us consider the DC operation. From the pullup-pulldown symmetry, even-numbered node ($2n$) voltage is $V_{2n} = (V_{DD}/2) + \Delta V$ and the odd-numbered node ($2n\pm 1$) voltage is $V_{2n\pm 1} = (V_{DD}/2) - \Delta V$.

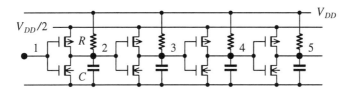

Figure 2.19.1 Phase transition in a gate-field

Let us compute the electrical power per stage consumed by an infinitely long gate-field:

$$P = G_m[(V_{DD}/2) - \Delta V] V_{DD} + (\Delta V/R)(V_{DD}/2) \quad (2.19.1)$$
$$= (G_m V_{DD}^2/2) + V_{DD}[(1 - 2G_m R)/2R] \Delta V$$

where the first term is the power supplied by the V_{DD} volt-power supply and the second term from the $V_{DD}/2$ volt-power supply of the resistor. For a steady state to be established, the entropy production must be the minimum. The power P minimum is reached at $\Delta V = 0$ if $2G_m R < 1$ and at $\Delta V = V_{DD}/2$ if $2G_m R > 1$ (I note here that the choice of an even- or odd-numbered index is independent of the physics in the indefinitely long chain, so $\Delta V = -V_{DD}/2$ is the minimum as well). The gate-field goes through a phase transition at $G_m R = 1/2$, and the uniform field or the alternating-field pattern is the minimum entropy configuration for the two different parameter domains. This is indeed a thermodynamical phase transition. For $G_m R > 1/2$ there are two stable states: the one the even-numbered nodes are HIGH, and in the other they are LOW.

In Fig.2.19.2(a)-(c) the input node at the left end is pulled up at time $t = 0$ by the stefunction having amplitude ΔV_0, and the node i voltage is written as

$$V_i = (V_{DD}/2) + \Delta V_i$$

and $|\Delta V_i|$ (the absolute value) versus node index i was plotted. This is to avoid clutter in the figure. The voltage profiles at different times are shown.

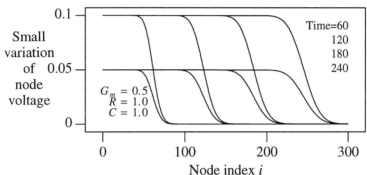

Figure 2.19.2(a) Small-signal wave propagation in a gate-field

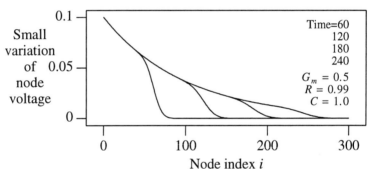

Figure 2.19.2(b) Small-signal wave propagation in a gate-field

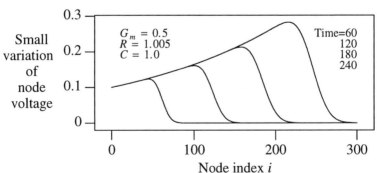

Figure 2.19.2(c) Small-signal wave propagation in a gate-field

Quantum-Mechanical Gate Field 181

In Fig.2.19.2(a) $G_m R = 0.5$. It shows propagation of two small amplitude stepfunctions having step height 0.1 and 0.05. The both small-amplitude step fronts propagate to the right at a constant speed of about one stage per unit time, which is determined by the gate-field parameter, independent of amplitude and waveform. The stepfunction transition spreads over about forty stages. By superposing many stepfunctions a propagating wave having any voltage profile whose transition edge is less steep than the natural stepfunction edge can be synthesized. Since the velocity is independent of the amplitude, the initial field profile maintains itself. This wave has no dispersion. This is very much like the electromagnetic wave, especially in the limit as $G_m/C \to \infty$. This is the limit we are interested in.

If $G_m R < 0.5$, the stepfunction input voltage attenuates as it propagates the gate-field, as shown in Fig.2.19.2(b). The amplitude decreases by factor $2G_m R = 0.99$ per stage, thereby creating the envelope of the gate-field potential profile. The envelope $\Delta V(x) = |V_i|$ is described by an equation

$$\frac{d\Delta V(x)}{dx} = [(2G_m R) - 1]\Delta V(x) \qquad (2.19.2)$$

where instead of the discrete-node index i we used a continuous-space variable x to indicate the location of the gate-field. The characteristic length of decay or the range is $|1/(2G_m R - 1)| = 100$. Different from the case of $2G_m R = 1$ in Fig.2.19.2(a), the gate-field *screens* the input effects. The input-voltage variation does not penetrate beyond the range. The screening effect finds a counterpart in the electroweak, theory, ultimately creating a gauge Boson having mass. If $G_m R > 0.5$, the stepfunction input voltage is amplified as it propagates down the gate-field as shown in Fig.2.12.2(c). This is, of course, because the gate-field has the hidden energy supply. Obviously, the increase in the amplitude does not continue indefinitely: if $\Delta V(x)$ reaches $V_{DD}/2$, the amplitude is clamped by saturation. The clamping takes place even if $G_m R$ is only infinitesimally larger than 0.5. Then $G_m R = 0.5$ is indeed a quite special condition. This special condition provides a wave in the gate-field associated with a massless, photonlike particle having an infinite range.

The concept of phase transition is relevant to a question if a gauge Boson has mass by the elementary particle theory. A gauge-field-like electromagnetic field is associated with an elementary particlelike photon that has no mass. The mathematical framework that leads to emergence of mass is summarized as follows (Aitchison, 1982). The following condensed summary is provided only to bring the beauty of the mathematical theory into perspective, and therefore it lacks completeness and precision. The 4-potential $\vec{A}$ of electromagnetic field satisfies the equation

$$\Box \vec{A} - (\text{gauge term with } \vec{A}) = \vec{J} \qquad (2.19.3)$$

if the variables that are conventional in the elementary particle theory are used. $\vec{J}$ is the 4-current (combination of the charge and the current densities), where the square symbol is the D'Alembert operator $\partial^2/\partial t^2 - \nabla^2$.

Equation of a field $\vec{A}$ that is associated with an elementary particle having mass is derived from the Einstein's mass-energy relation $E^2 = (P^2/2m) + (mc^2)^2$ by substituting E and P by the operators $jh(\partial/\partial t)$ and $-\nabla$, respectively, for quantization, and it has a form

$$(\Box + M^2)\vec{A} = 0 \qquad (2.19.4)$$

In the electroweak, theory a new field, a Higgs field ϕ, is introduced in the physical-vacuum background, and that provides a current density of Eq.(2.19.3) of the form

$$\vec{J} = \text{const.}[\phi^*(\nabla\phi) - (\nabla\phi^*)\phi] - \text{const.}|\phi|^2\vec{A} \qquad (2.19.5)$$

to the right side of Eq.(2.19.3). It is possible to erase the first term of $\vec{J}$ by sacrificing one degree of freedom in the gauge term of Eq.(2.19.3). We then obtain the equation of the required form, Eq.(2.19.4) and $\vec{A}$ represents the field associated with a particle having mass.

As discussed by Aitchison (1982), the essential physics is preserved by dropping the time dependence. Then the issue is how to make the two equations

$$\nabla^2\vec{A} = -\vec{J} \quad \text{and} \quad \nabla^2\vec{A} = M^2\vec{A} \qquad (2.19.6)$$

consistent. This is possible only by introducing a self-consistent current $\vec{J} = -M^2\vec{A}$. This term creates a screening effect. The field $\vec{A}$ depends on a distance parameter x like $\vec{A} = \vec{A}_0\exp[\pm x/(1/M)]$, and $1/M$ is the characteristic length. This leads to the physical conclusion that the weak, force mediated by the intermediate-vector Boson is a short distance force.

If we define the small deviation of the gate-field node voltage by

$$\text{Even node: } \Delta V(x) = V(x) - (V_{DD}/2) \qquad (2.19.7)$$

$$\text{Odd node: } \Delta V(x) = (V_{DD}/2) - V(x)$$

the field variable $\Delta(x)$ satisfies the wave equation of the form

$$\frac{\partial \Delta V(x,t)}{\partial t} + \frac{2G_m}{C}\frac{\partial \Delta V(x,t)}{\partial x} - \mu \Delta V(x,t) = \frac{J}{C} \qquad (2.19.8)$$

where

$$\mu = \frac{2G_m R - 1}{CR}$$

This equation has a general solution of the form

$$\Delta V(x) = \exp(\kappa x)F(x - vt) \quad \text{where} \qquad (2.19.9)$$

$$v = 2G_m/C \quad \kappa = \mu/2G_m$$

Equation (2.19.8) is a first-rank equation, reflecting the unidirectionality of signal propagation in the gate-field. The rank difference has quite reasonable explanation, and therefore it does not matter in the following discussion. The right-side term J/C is the external current injected to the node at location x divided by capacitance C, and that has the same meaning as the 4-current of Eq.(2.19.3). Except for the rank difference, Eqs.(2.19.4) and (2.19.9) are mathematically similar, and μ is the parameter that play the role of mass in Eq.(2.19.4). In Eq.(2.19.8) the right-side term J/C and the third term on the left side, $\mu \Delta V(x,t)$, play the same role, and this is clear if we observe Eqs.(2.19.3) through (2.19.6). The counterpart of the current supplied by the Higgs field already exists in the gate-field, since the gate-field has the background mechanism. Then μ is the parameter that represents the mass. Then we may identify that μ is the essential factor contained in the expression of *mass* of digital excitation.

From this viewpoint, the propagation of the attenuating wave in the gate-field when $G_m R < 0.5$ is quite similar to the propagation of an intermediate-vector Boson having mass. In this gate-field the coefficient of the term proportional to ΔV in Eq.(2.19.1), $V_{DD}[(1-2G_m R)/2R]$, is the parameter that determines the minimum of the entropy production P. Therefore, if that is zero by choosing R, the gate-field has high symmetry. By breaking the symmetry an attenuating wave that may be considered as the counterpart of a particle having mass is created. The mechanism of symmetry breakdown is supported by the existence of the structure behind the observable field, hidden from the observer. The hidden mechanism is equivalent to the Higgs field in the electroweak, theory. Then we have both physical and mathematical parallels between the two models.

As for the case of $G_m R > 0.5$, there could be two possible interpretations. The first is that as the amplitude of the wave increases, it saturates at $\Delta V = \pm V_{DD}/2$. At this point, the critical gain condition $G_m R = 0.5$ is effectively established by including the nonlinearity of the circuit, and it maintains the propagation of a constant amplitude wave. Is this a photonlike object? Although this is intuitively rational, the field is unable to carry an arbitrary waveform, and that is a fundamental character of the electromagnetic wave. The second interpretation is to focus attention on the two independent digital wave types the one that makes a LOW to HIGH and the other that makes a HIGH to LOW transition at a given node. Waves like this have features of a particle and an antiparticle. We observed in Sections 2.17 and 2.18 that in a long-gate field a pair of the two transitions collapses after a long time. From this viewpoint they are like quark and antiquark. They have mass, and if isolated, they carry apparent infinite energy (Section 1.04). I am inclined to support the second interpretation. If this view holds, a parallel to the unification of the three forces, the strong, weak, and electromagnetic forces, does exist in the gate-field model.

As the conclusion of this section, the problem stated at the beginning was answered only qualitatively. I was unable to arrive at an unequivocal mass formula of digital excitation. Yet we are successful in identifying a signal carried by a massless, photonlike particle and a signal that is carried by a particle like a mass point. In an analog circuit in its conventional mode of operation a massless particle like a photon carries information. In a digital circuit the signal is carried by a particle that has mass, like intermediate-vector Bosons and quarks. This is a new viewpoint that extends the horizon of the electronic-circuit theory. To the reader I wish to stress the following: to the question, "Why can such an apparently far-fetched conclusion be reached by electronic circuit theory?" there is a clear and undeniable answer already: because the electronic-circuit theory is the science of models and functions of physics. An equivalent circuit is able to represent anything that exists in the universe, and circuit theory is able to understand the equivalent circuit.

2.20 The Dynamics of Digital Excitation in Closed Path

In Chapters 1 and 2 I examined the motion of digital excitation in a linear one-dimensional gate-field. The motion is characterized by the velocity and, if we do rely on the primitive concept of the mass associated with it, by the momentum. Mechanics has one more basic parameter, the angular momentum, that characterizes a rotational motion of an object. As a close analogy to this type of motion, I introduce the motion of digital excitation in a closed path. The objective is to explore the physical meanings of the model and of the motion. More detailed study is postponed to Chapter 3, where some useful applications of the structure are discussed along with the theory.

A closed digital-signal path has a counterpart in the two conspicuous quantum-mechanical systems. The one is the electron trapped in the Coulomb force field of a positively charged atomic nucleus, and the other is the elementary particle itself, to explain its self-rotation. I discuss the model of the atom in Chapter 3, and here I explore the possibility of modeling the intrinsic angular momentum of an elementary particle, called *spin*. The spin of an elementary particle emerges as the consequence of idealization: if a rigid body that has nonzero size, whose position is described by six coordinates in the space, is shrunk to a mass point by idealization, its position is described only by three space coordinates. A spin appears to represent the three missing coordinates, describing its rotation.

The spin of an elementary particle has a profound effect on the statistics, when a many-particle system is built from the same kind of particles. Particles that have spin 1/2, 3/2, 5/2, . . . , are called Fermions, and particles that have spin 0, 1, 2, . . . , are called Bosons. Two Fermions in a system cannot occupy the same quantum state including the spin, that is, if two Fermions are otherwise in the same state and they have spin 1/2, one has spin quantum number +1/2 and the other -1/2: including the spin quantum number, the two Fermions cannot be in the same state. This restriction does not exist for Bosons. Since photons are Bosons, many photons can be in the same state, thereby increasing the electric field amplitude and create a strong light beam.

Exactly how the quantized spin is generated for each elementary particle type is a fundamental problem of elementary particle physics that has not yet been settled. One of the currently popular model that explains the origin of the elementary particles and their spin is the *string model*. My objective in this section is to show that the gate-field model is able to simulate the string model to a very significant extent. This is the reason that this section follows the previous section, which deals with the another current subject of elementary particle physics, the spontaneous breakdown of symmetry leading to the emergence of mass.

Since spin has its origin in a rotational motion of a particle, it is natural to model the rotational motion in a closed path of cascaded gates, shown in Figs.2.20.1(a) and (b).

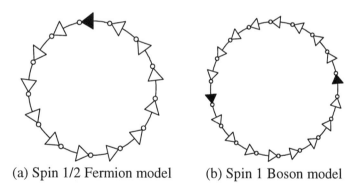

(a) Spin 1/2 Fermion model   (b) Spin 1 Boson model

Figure 2.20.1 Closed-string models of Fermion and Boson

In Figure 2.20.1(a) a thirteen inverter loop is shown. At the location of the inverter shown by the dark triangle, the rule of stability of an inverter chain, the alternate HIGH and LOW Boolean-levels, is broken. The location where the rule is broken moves around the loop. If the location turns around the loop twice, the circuit returns to the same state. It may be said that the loop contains half of the complete spatial wavelengths. In Fig.2.20.1(b) an eighteen inverter loop is shown. The rule of alternate HIGH and LOW Boolean-levels is broken at two places, shown by two dark triangle inverters. The loop returns to the same state if the locations turn around once or the loop contains one complete wavelength. From this observation, it is natural to identify the circuit of Fig.2.20.1(a) as representing a Fermion having spin 1/2 and the circuit of Fig.2.20.1(b) a Boson having spin 1. An odd-numbered member closed loop has an odd number of locations where the HIGH and LOW alternating rule is violated. Then a large odd-numbered member loop sustains waves having 1/2, 3/2, 5/2, . . . , waves in it. The ground state contains 1/2 wave. These states represent, respectively, the state of the Fermion having 1/2, 3/2, 5/2, . . . , spin. The ground state of an even-numbered member loop satisfies the alternating HIGH and LOW rules everywhere. This is the ground state having spin 0. If there are 2, 4, 6, . . . , violations of the rule, the Boson represented by the ring has spin 1, 2, 3, . . . , and so on, respectively. The loop model explains the high-spin Fermion and Boson states. We consider in this model that the effects of pulse collapse by the initial condition, as discussed in Sections 2.17 and 2.18 do not take place. As we discuss in Section 3.08, there is a technique for securing pulse width for an indefinitely long time by using clock.

We note here that the inverter output nodes are not the model of an ordinary spatial point, as I considered before. The nodes are the internal discrete points of a string, whose equivalent circuit model is shown in Fig.2.20.1. The length of the string has been estimated at about 10^{-13} (cm) in earlier string theory, but in recent superstring theory the length is estimated at the much smaller 10^{-31} (cm) range, where even the dimensionality of physical space-time becomes the subject of controversy. My point is that as the science of models itself, modern electronic-circuit theory hopes to provide a model even in the extreme domain of physics.

The closed-string model of Fermions and Bosons would be of limited interest if it were able to explain the various spin states. If two Fermions stick together, a Boson is formed. The Boson has spins, either the sum or the difference of the spin of the two Fermions. Figure 2.20.2(a) shows two Fermions having spin 1/2 in a nine-member loop approaching together. As the loops touch, they connect together, thereby forming a single loop shown in Fig.2.20.2(b), giving spin 1. In the original two loops of Fermions, the locations of viloation of the alternating HIGH and LOW rules are far apart. When the loops join to form an even eighteen member loop, there are two locations where the rule is violated.

If the violations of the alternating rule of the original loops occur at the inverters that are immediately at the left and at the right of the joint, A or B, the two violations cancel, and the resultant eighteen-member Boson loop has no location of the rule violation. This is the formation of a Boson having zero spin. In this way the model is able to explain why a combination of two Fermions makes a Boson. Similarly, it is shown that combination of two Bosons creates a Boson.

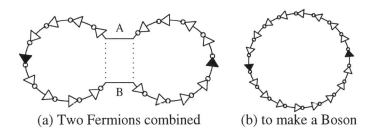

(a) Two Fermions combined (b) to make a Boson

Figure 2.20.2 Two combined Fermions make a Boson

Before concluding this chapter, I must repeat the following: to consider a mass of digital excitation, and to make a model of a fundamental particles of nature by an electronic-circuit model, might appear an extreme stretch of imagination. But electronic-circuit theory is the only branch of physical science that addresses the model itself. As long as physics requires a model, the merit of the electronic-circuit theory that it allows easy model building and scrutinization cannot be dismissed. Indeed, a digital-circuit theory might become the most convenient medium for creating the most fundamental model of nature. One such aspect is that circuit theory is spatial-dimension independent. The second reason for this investigation is many new ideas that benefit state-of-the-art circuit design emerge naturally from the search for parallels between the models of two areas that previously had been considered unrelated.

Chapter 3

The Macrodynamics of Digital Excitation

3.01 Introduction

In the preceding chapters I discussed switching transients in the gate-field and discussed how they are similar to and different from the quantum-mechanical phenomena of the microscopic world. In this chapter I discuss why macroscopic objects like a logic gate and a digital excitation exhibit features that are substantially quantum-mechanical. This attempt includes a new feature. Instead of identifying an approximately localized excitation in a classical particle, we consider a state extending over the entire circuit. For an entire circuit to have a simple, integrated state, the circuit must have a means of integration, which is feedback. We consider the closed loops of a cascaded circuit or generally, a multiple-loop ringoscillator. To carry out this program, we need a significantly different mathematical approach, not at the circuit level but at the gate level. This method has an interesting feature: It allows interpretation of the digital circuit's quantum-mechanical phenomena in terms of classical mechanics, thereby allowing easy understanding, but simultaneously making the boundary between the two regimes quite unclear.

Before going into the subject we need to clarify the attributes of quantum-mechanical systems. Quantum mechanics in its present perfected form derives the attributes of quantum-mechanical objects by solving basic equations, such as Schroedinger's wave equation or Heisenberg's matrix equation. There are certain rules for applying the mathematics to a given system. For investigating nature at its microscopic level, this is adequate. Yet there are other objects that are artificially made by humans that have many similarities to the microscopic objects described by quantum mechanics, and yet are not governed by the classical equations or concepts. To answer the question, "What characterizes a quantum-mechanical object?" we need to consider many qualitative attributes of the object.

This question is not simple. To begin with, let us consider what a classical object and a quantum object are. A particle is a small object in the classical interpretation, whose location and momentum can be measured at any time without perturbing its state of motion. In quantum mechanics, a particle may not be a small object relative to the entire system size being considered. Loss of the attribute of small size by the classical particle is the characteristic feature of a quantum-mechanical object. An example is an electron cloud around an atomic nucleus. A small electron becomes an extensive and complicated electron cloud around the nucleus. Whether the particle is a small object or a spatially extended large object depends on the method of observation. A direct manifestation of this ambiguity is the impossibility of determining its location and momentum simultaneously. A simple small particle becomes a complex system like object, which is one distinctive feature of quantum mechanics.

In classical mechanics, energy and matter are different objects. In a relativistic quantum mechanics, mass is considered to be a form of the existence of energy. Thus, energy can form matter, and matter can be converted back to energy. In the digital-gate field, we have this point of view already taken into consideration. In this regard, the gate-field model is quantum-mechanical.

A particle is an individually identifyable object in classical mechanics. In spite of its small size, it carries a tag, by which its absolute identity can be maintained all the time. From a fundamental viewpoint this is a contradiction. A particle loses its identity in quantum mechanics, and this is considered more rational than the assumption of classical mechanics. Two elementary particles like electrons cannot be distinguished if they occupy the same space, such as a potential well. Metaphorically, an elementary particle is so *elementary* that no inscription of its identification can be engraved on it. Indistinguishability is just as fundamental as the particle-wave duality.

The issue of distinguishability is more involved, since certain classical objects may be or may *become* indistinguishable. The common currency is used on the assumption of indistinguishability. Army commanders and certain company executives may think that their soldiers and employees are indistinguishable, characterless particles, as well. In classical mechanics, head-on colliding particles may fuse together and lose their identity and kinetic energy. The energy lost to the internal degree of freedom is unrecoverable as macroscopic energy. This means that classical mechanics is incomplete: it must be supplemented by thermodynamics and solid-state physics to resolve the issue. The quantum electro- and chromo-dynamics are complete or self-contained in this sense. This issue creates a certain ambiguity even in classical mechanics. Is the indistinguishability quantum-mechanical? The question is essentially depends on the viewpoint. I consider that it belongs to quantum-mechanics.

Let us study a quantum-mechanical system having small size, such as an atom. The simplicity allows clear interpretation. The subject of this chapter's study a digital circuit that models a small quantum-mechanical system as a whole has the following fundamental difference from the excitations in the extended gate-field discussed in the first two chapters. The gate-field was a model of physical space, and the excitations were a classical, as well as a quantum-mechanical, object. In a small digital circuit modeling a simple quantum-mechanical system, the circuit itself is a quantum-mechanical object. The duality is carried over from quantum mechanics itself. While quantum mechanics is used to explain the atomic spectra, it is applied on a simple object like an atom, and the object itself is quantum-mechanical. If it is applied to study creation and annihilation of elementary particles in physical space, the excitation becomes the quantum-mechanical object. Yet on the other extrema, an excitation that travels in a gate-field is essentially like a classical particle. We need to study various aspects of the cases.

3.02 Quantum States

An attribute of a quantum-mechanical object is that it is in one of generally many *eigenstates* that are determined by the structure of the object, including its structural symmetry. In the truly elementary microscopic object like atoms, the symmetry is absolute and decisive in determining the nature of the state. Some states are stable, and others are not. Generally, a state having higher symmetry is more stable than a state having lower symmetry. Some states maintain themselves as long as the object is not severely disturbed. Some other states have finite lifetimes, and they decay to one of the more stable states. What are the electronic equivalents of objects that have such character? It is possible to build two types of circuit models, depending on the criteria: the state is observed either statically or dynamically. Static observation is to connect two terminals of a test equipment like a direct current (DC) voltmeter directly to a pair of the nodes and to determine the node voltage. In a dynamic observation the equipment detects the existence or nonexistence of a time-dependent voltage variation. Static observation is for a particle in a gate-field. Dynamic observation is for a wave sustained by the entire system.

A circuit that has multiple states for static observation is a latch, shown in Fig.3.02.1(a). The latch stays in a state as long as it is powered. The state that the latch is in was established before, often at the time when the power is turned on. A piece of quantum-mechanical test equipment has a strong drive capability to read a node voltage and restore it to the ideal level if any level deterioration should exist. In the latch circuit of Fig.3.02.1(a), the restoration capability exists in the circuit itself. A transition from one state to the other is effected if the latch is subjected to a strong external drive and a temporarily breaking of the measure of systemwide integration, the feedback. As the state changes, an apparent transfer of energy from one node to the other takes place. The transition involves apparent charge transfer among the nodes of the circuit (Section 1.04). Such a transfer creates a time-dependent dipolar moment that emits electromagnetic radiation to the space. The circuit model has many similarities to a light-emitting transition of an atom. Yet the properties of a static latch is more closely similar to a quantum-mechanical object like spin, whose internal state is not accessible for observation. Element of electron charge is an another example. The similarity between a binary latch and the spinlike quantum-mechanical state is obvious. What I wish to show in this section is that other, *electron – in – atom* like objects exist in the digital-gate model.

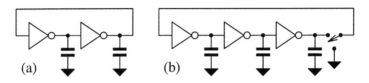

Figure 3.02.1 A circuit model of a
two-state quantum-mechanical object

A circuit that has quantum-mechanical properties to a dynamic observation is shown in Fig.3.02.1(b). This is a circuit called a *ringoscillator*, that can be fitted with an additional control circuit to start-stop oscillation. If the circuit is not oscillating, the circuit stays in a quiescent state. If it starts oscillating, the dynamic state sustains by itself. The dynamic representation of binary states has practically never been used in conventional electronics, but it is used to construct logic circuits of an animal's central nervous system. The dynamic representation of a state may have several variations. Dynamic representation may use oscillation/no oscillation, oscillation amplitude, or oscillation phase as the variable to indicate the states. Strictly speaking, the dynamic state loses its energy by radiation as the classical model of an atom does, but in the circuit model the lost energy is made up by the power supply in the background, and therefore a model very similar to an atom in the quantum state can be built. In the binary dynamic representation, a state can be established only when the waiting time of the order of the period of oscillation after the control to start or stop oscillation is exercised. During that time the circuit has no Boolean-level: determination of the Boolean state of the dynamic representation requires time of the order of the oscillation period. Although the static representation of a state has a similar problem originating from the circuit's delay, the time is generally shorter. There are features of a ringoscillator that make this object look more like a quantum-mechanical atom. Let us study some of the pertinent features.

A ringoscillator may be considered as a macroscopic object, within which a well-defined elementary excitation moves from one node to the next. This is indeed the case for a ringoscillator consisting of a long closed chain of gates, operating at the fundamental frequency. The node waveforms and the Boolean states of the seven nodes of a seven-stage ringoscillator made from identical inverters are shown in Fig.3.02.2. The FETs are modeled by the gradual-channel, low-field model. The inverter parameter values are chosen as follows:

$$V_{DD} = 1.0 \quad V_{THN} = V_{THP} = 0.1 \quad B_N = B_P = 1.0 \quad C = 1.0$$

The open and closed bars above the node waveforms indicate HIGH (closed bars) and LOW (open bars) states of the nodes. The top row of the state bar is for the dotted curve in the node-voltage waveform plot. Between the time domains having definite Boolean-levels, there is an interval where the quantum-mechanical node-voltage measurement of Section 2.09 gives only a probabilistic answer.

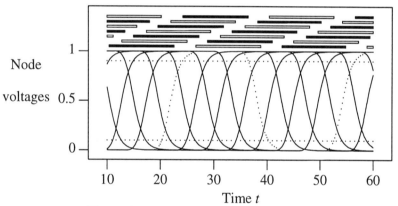

Figure 3.02.2 Ringoscillator node-voltage waveforms and the Boolean-level

If the number of stages increases, the fraction of the region where the Boolean-level is indefinite decreases. In the limit of an infinitely long ringoscillator, the Boolean-levels of all the nodes are definite all the time, and the ringoscillator is explainable by the classical mechanics of circulating excitation as a classical particle. In Fig.3.02.2 the gap between the open and closed bars that indicates the transition region of the Boolean states moves with time. This is the motion of the *classical* particle around the loop. This is an example of a general principle that a quantum system in certain limits of large scale behaves like a classical mechanical object. This principle, Bohr's correspondence principle, is discussed in the next section from a different angle. Here I point out the following perspective: a ringoscillator having many inverters is able to support many modes of oscillation, as we see in Sections 3.08 and 3.09, and the classical operation we saw is only one of the modes having the lowest oscillation frequency. The smaller-scale ringoscillators operate successively more in quantum-mechanical mode.

If the ring is short, and if the inverter parameter values chosen are not uniform, a ringoscillator behaves differently from the classical mode. Figure 3.02.3 shows the node waveforms and the Boolean-levels of a three-stage ringoscillator.

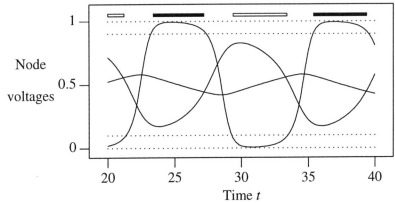

Figure 3.02.3 Ringoscillator node-voltage waveforms and the Boolean-level

The FETs are modeled in the same way, and the long-channel, low-field model parameter values are selected as follows:

$$V_{DD} = 1.0 \quad V_{THN} = V_{THP} = 0.1 \quad C = 1.0$$
$$B_{N1} = B_{P1} = 0.1 \quad B_{N2} = B_{P2} = 4.0 \quad B_{N3} = B_{P3} = 4.0$$

Only node 3 reaches the definite Boolean-levels, and still there are wide time zones where the Boolean-levels are uncertain. As for nodes 1 and 2, they never reach any definite Boolean-level. In this oscillator the question, " Where is the digital excitation?" cannot be answered, as it is impossible to determine the location of an electron in the Bohr orbit on an atom, and therefore a classical interpretation fails. The entire structure operates as a single, integrated quantum-mechanical object. This observation gives an interesting idea: a quantum-mechanical object is indeed a very well-integrated object, but a classical object can be assembled from the independent parts. Classical mechanics reflects reductionism thinking; quantum mechanics does not.

We are able to show that the extreme example of a quantum-mechanical ringoscillator is a conventional phase-shift oscillator, consisting of a single inverting amplifier and two stages of cascaded RC low-pass filters, shown in Fig.3.02.4.

Macrodynamics of Gate Field

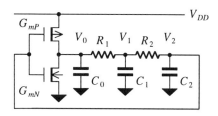

Figure 3.02.4 A phase-shift oscillator

Nodes 0, 1, and 2 are observable, since they are all capacitance-loaded. The node voltages V_0, V_1, and V_2 satisfy the circuit equation

$$C_0 \frac{dV_0}{dt} = I_P(V_0, V_2, V_{DD}) - I_N(V_0, V_2)$$

$$C_1 \frac{dV_1}{dt} = \frac{V_0 - V_1}{R_1} - \frac{V_1 - V_2}{R_2} \qquad C_2 \frac{dV_2}{dt} = \frac{V_1 - V_2}{R_2}$$

where I_P and I_N, the PFET and the NFET currents, respectively, are given by the gradual-channel, low-field FET model, and the numerical solutions at the steady-state oscillation are shown in Fig.3.02.5. The parameter values are given in the figure. No node swings enough high or low voltages to provide a definite Boolean-level determination. The phase-shift oscillator of Fig.3.02.4 is often used to generate nearly sinusoidal waveform. This means, in our present interpretation, that the circuit does not go deeply into the saturation region and it is truly quantum-mechanical. We see the nature of saturation effect: saturation creates digitized levels of a latch, but nonsaturation of node voltages creates a quantum-mechanical ringoscillator.

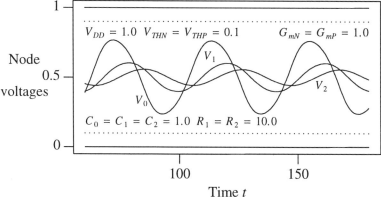

Figure 3.02.5 A phase-shift oscillator as a quantum-mechanical object

In this section we showed that a latch and a ringoscillator can both be quantum-mechanical objects, but at a large size a ringoscillator approaches a classical object. Let us investigate this interesting aspect of the effects of the size of the system on the type of mechanics in the next section.

3.03 Bohr's Correspondence Principle

Quantum-mechanical states exist at the most countable infinity in number in a finite-size system, and an integer index can be assigned to each state. Among the states there is a state that has the minimum energy. Starting from the ground state, all the states can be arranged in the order of increasing energy, by an increasing integer index n. Then the states having high indices, or the states in the limit as $n \to \infty$, present the object in a classical-mechanical mode of operation (Landau and Lifshits, 1958). This is Bohr's correspondence principle, which connects quantum and classical physics. An example of the principle is the electron state having a high principal quantum number in a hydrogen atom. What would be the circuit-theoretical counterpart of this principle? We saw in the last section that a ringoscillator approaches the classical operation in the limit as long ring and as low oscillating frequency.

The quantum-mechanical principle implies that the energy difference between two consecutive states diminishes as index n increases, as happens in the case of a hydrogen atom. This is because the energy is bounded, and if there are an infinitely large number of energy levels, the set of the energy values must have an accumulation point. Close to the accumulation point the gap between the two consecutive energy levels tends to zero. In the limit as $n \to \infty$, a state can be found practically for any arbitrary given energy, and energy quantization becomes invisible. This is not the case for small n, where quantum mechanics gives qualitatively different results from classical mechanics. What is crucial here is to have many energy states that make the energy appear continuous.

A prerequisite to studying the meanings of Bohr's correspondence principle in the static latch circuit is to construct a circuit that has an arbitrarily large number of static internal states N and to show that the circuit behaves like a classical object in the limit as $N \to \infty$. Just an assembly of many binary latches [like at the input of a digital to analog (D/A) converter] does not provide a proper example, since voltage discretization still exists. The latch circuit must have any desired number of discretized states, each having its own quantized voltage levels. A latch built by applying positive feedback to an N-ary multilevel buffer shown in Fig.3.03.1 has N discrete states. The N-ary buffer works as follows. The N power-supply voltage levels $V_{S1}, V_{S2}, \ldots, V_{SN}$ satisfy $V_{S1} < V_{S2} < \cdots < V_{SN}$. The buffer symbols show conventional CMOS inverters, numbered by index i, that range $1 \le i \le N-1$. The i-th inverter is connected between power supplies V_{Si+1} and V_{Si}. The switching threshold voltage of the CMOS inverter is V_{GSW}, referred to its negative power-supply. If input voltage V is in the range $V_{Si} < V < V_{Si} + V_{GSW}$, the i-th inverter output is at V_{Si+1}. For any j ($< i$), the output voltage of the j-th inverter is V_{Sj}, and for any j ($> i$), the output voltage of the j-th inverter is V_{Sj+1}. The output of the N-ary buffer is driven by a series connection of a PFET and an NFET. The NFET and PFET at location i are both conducting, and they connect the power supply V_{Si} to the output. All the NFETs in the location j ($< i$) are off, and all the PFETs in the location j ($> i$) are off, as well. Then all the power supplies V_{Sj} ($j \neq i$) are disconnected from the output. The output is driven to voltage V_{Si}. The case $V_{si} + V_{GSW} < V < V_{Si+1}$ can be understood in the same way, and in this case the output is connected to power supply V_{Si+1}. If this buffer is used in a positive feedback loop to build a latch, the latch circuit stays in one of the states to which it was set at the beginning. The N-ary buffer requires N power-supply voltage levels. The output voltage of the N-states is one of the N power-supply voltages $V_{S1}, V_{S2}, \ldots, V_{SN}$, where $V_{S1} < V_{S2} < \cdots < V_{SN}$. By triggering the latch, a transition from any state to any other state can be effected. The output voltage of the buffer is developed across capacitance C, which is connected between the output node and the lowest power supply V_{S1}. The range is limited between the highest- and the lowest-power supply voltages. If the maximum voltage range available, $\Delta V_P = V_{SN} - V_{S1}$, is fixed, the output-node voltage differences between two consecutive states decrease with increasing N. In the limit as $N \to \infty$, the buffer becomes a continuous analog buffer having unity gain all over the voltage range.

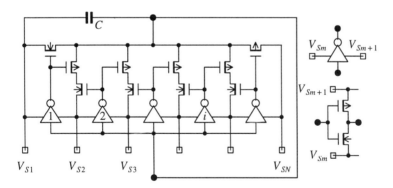

Figure 3.03.1 A circuit model of an N-state quantum-mechanical object

In this limit it is still possible to design the circuit such that $\Delta V_{P2} = V_{S2} - V_{S1}$ remains the same, but the rest, $\Delta V_{Pm} = V_{Sm} - V_{Sm-1}$ ($m \geq 2$), decreases with increasing N. Then the circuit model is very similar to an atom, which has a single deep ground state and many closely spaced excited states at the high energies. The latch works as an analog voltage storage circuit for $m \geq 2$, although the accuracy is not high enough for practical use. The buffer is indeed a hybrid of a digital and an analog buffer. This circuit model shows that an analog circuit that operates using continuous values of voltage variable is a classical-mechanical circuit model, and a digital N-ary buffer circuit using discrete voltage values is a quantum-mechanical model, and there are many buffers whose degree of discretization is between the two extrema. The N-ary buffer and latch have not been used practically, but there is a subsystem-level circuit similar to that, which is used very widely in the telephone system. A μ-law CODEC used to encode and decode the telephone voice signal to PCM data stream works effectively as a continuous buffer for small-amplitude voice signal, but as a discrete buffer at a large amplitude. The subsystem is indeed a close model of a quantum-mechanical atom.

A more than two-level logic gate has many similarities to two-level Boolean logic gate. Figures 3.03.2(a) and (b) show the input-voltage and output-voltage relationships of a four-level logic gate. In the following studies the gradual-channel, low-field FET model not including the back-bias effect in the figures, with the set of parameter values, is used. The four logic-level voltages are 0, 1, 2, and 3. All the node capacitances are set at 0.1. The other parameter values are given in the figures.

Figure 3.03.2(a) shows the input-output characteristic at the high input-voltage rise-fall rates. The characteristic shows a hysteresis, which is also observed in the conventional Boolean inverters if it is switched fast. Hysteresis is an indication of a dynamic switching process. If the rise-fall rates are reduced, the hysteresis practically vanishes, as is shown in Fig.3.03.2(b).

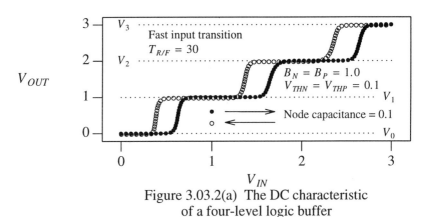

Figure 3.03.2(a) The DC characteristic of a four-level logic buffer

Macrodynamics of Gate Field

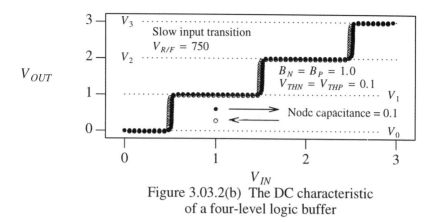

Figure 3.03.2(b) The DC characteristic of a four-level logic buffer

Then the common switching-threshold voltages between the two consecutive logic-voltage levels approach the center between the two voltage levels, since each binary inverter is pullup-pulldown symmetrical.

Although the input-output characteristic of the four-level gate is similar to that of a Boolean inverter, the similarity does not extend to the dynamic switching characteristics, and here is an interesting detail. The characteristics of Figs.3.03.2(a) and (b) are determined for a single gate, by varying the input voltage continuously all the way from V_0 to V_3 and then back to V_0. If many four-level gates are cascaded to make a gate-field, the signal propagation characteristics are quite complex. This is one of the practical deterrents of using more than two-level logic gates in an extensive logic system. Figure 3.03.3(a) shows propagation of the upgoing stepfunction at the input from V_0 to V_1 (closed circles), from V_1 to V_2 (open circles) and from V_2 to V_3 (× symbols). The parameter values are given in the figure. Figure 3.03.3(b) shows propagation of the downgoing stepfunction at the input.

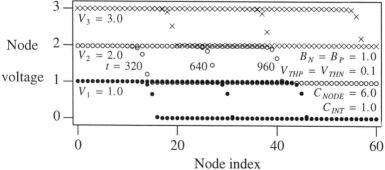

Figure 3.03.3(a) Unit step-wave propagation at three levels

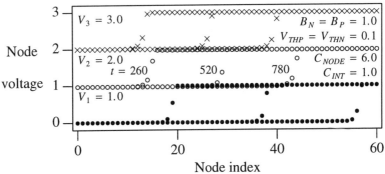

Figure 3.03.3(b) Unit step-wave propagation at three levels

In the two figures we note that the propagation of the stepfunction at different levels has different velocity. The upgoing and downgoing stepfunctions propagate at different velocities. The propagation velocity of an upgoing pulse from V_2 to V_3 is the fastest of the three upgoing stepfunctions, and that of the stepfunction from V_1 to V_0 is the fastest of the downgoing stepfunctions. This is understandable from the structure of the gate circuit. The switching of one of the inverters at the input is immediately relayed to either single pullup or pulldown FET, and the output node voltage makes the transition in a short time.

The velocity difference results in a breakup of a stepfunction that makes more than one step of transition, such as from V_0 to V_3. Figure 3.03.4 shows that the stepfunction breaks up as it propagates through the gate-field. A part of the V_0 to V_3 stepfunction goes ahead of the rest of it, thereby making a two-step structure. V_1 to V_3 stepfunction is kept integrated all the time.

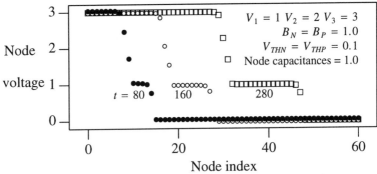

Figure 3.03.4 The breakup of excitation in a cascaded multilevel logic buffer field

This is a phenomenon very similar to the breakup of an unstable

Macrodynamics of Gate Field

pseudo particle into more stable final particles, observed in the elementary particle collision. The difference in the velocities of the stepfunctions depends on the gate-field parameter such as the FET characteristics and the node capacitances. If the parameter values are properly chosen, a multilevel stepfunction can be made stable. This observation gives an interesting general insight: what is stable and what is not depends on the field and its parameters and not on the nature of the excitation.

3.04 States of Nodes and Circuits

The state of a static latch node maintained by a positive feedback is in a strongly held static state. The state of a node in a gate-field driven by a static gate without positive feedback is in a normal static state. The substance of both types of state is the energy stored in the node capacitor, but the controllability of the energy is significantly different. In a strongly held static state, the energy held in the capacitor is defended actively by the latch circuit as long as the positive feedback continues to be applied. Since positive feedback maintains the state of the node, the state cannot be upset easily. Even if the positive feedback is removed, the node remains in the state for some time by the conducting FET and by the charge stored in the node capacitor. To control the state of a strongly held static node requires both a clock signal that turns the positive feedback on and off (or a strong and conflicting drive of the node) and the delay time to move the charge from the node capacitance. Controllability of a strongly held static node is small. In a normal static state, a state of a node can be changed by the input signal to the gate that controls the node. The charge and the energy appears to move from one node capacitor to the next node capacitor, as observed by the limited observer of the gate-field (Section 1.04). In the background, the energy exchange between the power supply and the node capacitors takes place. A moving state of the node looks like a static state if the observer moves with the excitation. From this viewpoint the two types of states are similar. In either case, their substance is the energy in the node capacitor. In a strongly held static node the capacitor is fixed in the space, and in a normal static node it moves at a certain velocity.

The mechanism behind the scene that effects the apparent motion of the charge or the energy consists of the three operations: charging capacitor at the second location, discharging capacitor at the first location, and information or signal transmission from the first to the second location. If we know the mechanism behind the scene, what is really moving is not energy but information. We observe, however, as a limited observer having nothing more than a logic probe, only the movement of energy, from which we infer the moving information. As long as the background circuit mechanism is hidden from the observer, the charge or the energy at one location appears to move to the next location. Strongly static and normal static states are different only in the apparent energy transfer mechanism in the background, but that is hidden from the observer if the logic probe does not disturb the node.

To compare the strongly held static and the normal static states clearly, we recognize that a logic gate is able to save information during the processing, and it is working as a memory as well. The two types of memories conventionally distinguish synchronous and asynchronous sequential circuits (Mano, 1991; Nagle, Carrol and Irwin, 1975). A latch works as a memory, and its strongly held static states have a direct counterpart in the stable quantum states. Similarly, the memory effect of a logic-gate circuit finds its counterpart in the wave function like quantum-mechanical states. The ringoscillation can be stable because of the *invisible* background energy-supply mechanism. The existence of such a stable state makes a ringoscillator look like an atom. The Bohr orbit is the closed loop of gates, on which electronlike excitation travels. This is the second view of a closed buffer chain, in addition to the model of spin in Section 2.21.

The memory effect of an inverting buffer during its switching is understood from Fig.3.04.1(a). A capacitance-loaded CMOS inverter, whose input voltage is less than V_{THN}, maintains its output stably at V_{DD}, thereby retaining the results of the preceding output pullup operation. The output-node state is conditionally stable for some time close to the Boolean-level it was set in, if the input voltage ranges $V_{THN} < V_1 < V_{GSW}$, if the gain is high. The retention of the state set by the last switching operation may be considered as an inertia on the circuit that provides the dynamic memory effect.

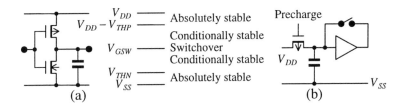

Figure 3.04.1 Static and dynamic memory

The memory effect is the mechanism that emerges in a large-scale circuit. In a system consisting of many gates like a gate-field, excitation moves from a gate to a gate, while that happens, memory effect is associated with it. An analogy to a crystal is appropriate. As long as the atoms at the lattice sites do not interact, all the properties displayed by the crystal are the properties of the atoms. The crystal displays no collective phenomena like electrical and thermal conduction, phase transition, or ferromagnetism. Propagating excitation like phonon and magnon is a manifestation of interaction between the atoms that retains the memory in solids.

It is necessary to clarify the terms. A state is a common concept in quantum mechanics and in digital-circuit theory. Quantum-mechanical eigenstates are the fundamental *modes* of the system that have no substructure. An elementary particle is an excited state of the physical vacuum. In digital-circuit theory, the same word *state* is used to represent the state of a system, not the circuit. Therefore a state has still two lower level substructures, a microstate (Shoji, 1992) and a submicrostate (Shoji, 1996). This may look as if comparing the two is not appropriate. A microstate of a digital circuit is defined by the set of specifications of the conducting states of all the FETs. An FET is in one of the three bias regions N (nonconducting), S (saturation), or T (triode) region. An FET in an N, S, or T region is equivalent to an open circuit, a controlled current generator, and a short circuit, respectively. Operation of a digital circuit can be described by the sequence of the microstate changes. In a steady state of a CMOS digital circuit, all the FETs are in any of the N or T states. They are steady states in the sense of the irreversible thermodynamics, since the states produce the minimum entropy (a CMOS circuit carries no standby power). An FET in an S state carries current, and the voltages of its source or drain nodes change. The state is not a steady state.

A state of a microprocessor is characterized by a set of Boolean-levels making a *state vector* characterizing the system's state, which are stored in the control latches. Control latches store the control data and are distinct from the data latches, which store the data to be processed. The control data are for executing the operational code. The logic circuits are driven in the next clock cycle by the control-latch data. State of a system is a specific concept, closely tied to its function. The state of the system and the microstate of the logic circuit are related as follows. Before the clock that executes the intention of the control data arrives, all the FETs of the logic circuit are either in an N or T state. If we try to establish a correspondence between the microstates of a logic circuit and the quantum-mechanical states, microstates as a whole do not have a direct counterpart to quantum mechanics. Only a subset of microstates finds proper counterpart. A microstate that includes at least one FET in an S state effect a transition between the steady states of the digital circuit. This is a state that is a probabilistic mixture of the steady states, which carry definite Boolean-levels of all the nodes. A microstate that has at least one FET in the S state is represented by a probabilistic superposition of the steady states. This is translated back to quantum mechanics as follows: a counterpart of the dynamic microstate that has at least one S state FET is a transitory state between the two quantum states. In quantum-mechanics, a transitory state is expressed as a superposition of the eigen states (Landau and Lifshits, 1958), with coefficients that determine the probabilities of finding the system in the superposed states, as it is practiced in the quantum-mechanical perturbation theory. The probabilities change with time. The transitory microstates are required to describe the behavior of a digital circuit as an analog circuit. In digital-circuit theory, a transitory state including S-state FET is used to avoid a cumbersome probabilistic superposition of the states, but the basic mathematics is the same as is used in time-dependent perturbation theory in quantum mechanics.

The relationship between the states and the microstates described in this section has one more detail. A microstate consisting only of N- and T- state FETs can still be classified into groups. If a node of the circuits is connected to a power supply bus through a conducting FET (in the T-region), the output nodes are held statically. If the node is loaded, energy can be drawn from the power supply through the conducting FET. If the node has a capacitor, but if its voltage is maintained by the charge stored before, and if the node is not connected to any power supply level by any FET in the T state, the node is dynamically held. The voltage is affected by the voltage measurement much more significantly in a dynamically held node than in a statically held node. An example of such a node is the precharged logic-gate output node shown in Fig.3.04.1(b) while the switch is open. Node voltage V is set at V_{DD} when the *Precharge* signal makes downgoing transition. On termination of precharge the node is isolated from the power supply, and V may decrease by leakage or by any other circuit imperfections. To determine the Boolean-level, the switch of the test equipment is turned on. To measure the Boolean-level, some charge must be drawn out of the capacitor. Then the circuit determines the Boolean-level stored as the capacitance charge, and after that is over, the observation process restores the correct Boolean-level voltage. This is the basic scheme used in an MOS DRAM: the memory read process *refreshes* the cell. The reading process of an MOS DRAM is essentially a quantum-mechanical observation process of a node held by the stored charge.

If a digital circuit includes a resistance, a microstate in which all the FETs are in N or T state may not be a steady state in a thermodynamical sense. If a node is reached only by a path including resistance, the node may not reach a steady state at the moment the last FET changed state from S to either N or T. An *exponential* tail of the transient is an unclean feature in digital-circuit theory, as I discussed in my previous work (Shoji, 1992). Yet the feature addresses an interesting question: what is the definition of a Boolean state in the time domain? The question is whether states established at different times approaching a steady state are the same. In digital-circuit engineering, this issue was circumvented by providing bands of voltage regions that correspond to the Boolean-levels, and this is a compromising engineering choice.

The other substructure of a microstate, a submicrostate, occurs only if the operation of a microstate including at least one S-state FET is analyzed. If this is done in a digital circuit including inductance, many interesting issues surface, as I discussed in my last book (Shoji, 1996).

The similarities that a digital circuit has to the other physical objects on one side to a quantum-mechanical microscopic object, as we discussed in the last section, and on the other to a macroscopic object, including the cooperative phenomena in the statistical physics as we discussed in Section 1.12 will have interesting consequences. A gate-field is the most accurate electronic model of the neuronic system of an animal brain. Then the operation of an animal brain will have two corresponding extreme modes of operation quantum-mechanical transition as well as classical and thermodynamic phase-transition. I suspect that the ultimate intelligence can be built only if we use hardware that has both features: the one provides flexibility, and the other provides reliability.

3.05 The Capability of a Circuit to Store Information

Let us consider an attribute of an electronic signal-processing circuit to gain some insight into an extensive circuit like a gate-field. An input signal goes into the circuit, and after a delay time the output signal is delivered. During the delay time, the information is retained in the circuit, while the signal travels from the input to the output. How much information is retained in the circuit and how? This is an interesting generalization of the issue related to the gate's temporal information-storage capability discussed in the last section. The information stored in the circuit is not in a digital format. It is stored as an analog charge or voltage variable in the node capacitor. The analog storages are dynamic storages. In an analog circuit, the sinusoid cycle portion that is being processed by the circuit is considered as stored in the circuit. The portion is often referred to as the phase lag. It appears that a lot more information is stored in a digital circuit than in an analog circuit because it has dedicated storage, the latches. Let us first consider how much information can be stored in a transient of a linear circuit. A simple example of a linear circuit is a cascaded RC chain that works as a low-pass filter. Let the left end of the RC chain shown in Fig.3.05.1 be driven by a sinusoidal voltage source, and we observe the voltage profile of the RC chain.

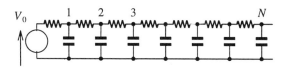

Figure 3.05.1 The memory capability of an RC chain circuit

Since a sinusoidal wave is characterized by periodic ups and downs, the memory capacity of an RC chain is characterized by how many ups and downs can be distinctly maintained in the RC chain voltage profile. In Fig.3.05.1, the input voltage is a biased sinusoid given by

$$V_0 = 0.5 V_{DD} [1 - \cos(0.1t)]$$

and the initial condition was $V_n = 0.5$ for all the $n \geq 1$ nodes. Figure 3.05.2 shows a numerical analysis result of a fifty-stage RC chain.

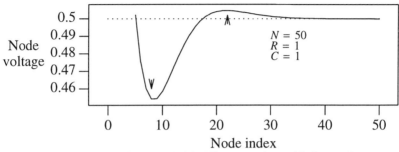

Figure 3.05.2 The retention of information in an RC chain

The parameter values are $R = 1$ and $C = 1$. It is interesting to note that the chain sustains only one period of the sine wave and no more. The conclusion is independent of the length of the chain or the frequency of the input signal. The analog memory capacity of an RC chain circuit is quite low. As for an LC chain, it is able to hold any volume of information in it, and the limit is set only by the discretization and the length. This is the fundamental difference between an RC and an LC line. The information fed into the RC line is quickly consumed and lost. Therefore, if the RC chain is to be used for signal processing, the processed signal must be taken out from the line before it decays, and it must be amplified again. The role of an amplifier is to counteract the signal decay. Then a chain of amplifiers retains much more information in it. If one stage of RC low-pass filter and an ideal amplifier are cascaded alternately, the capacity of the chain to store information increases. Since a real amplifier has nonzero internal resistance and output capacitance, it can be cascaded without an RC low-pass filter. This is the structure of a buffer field.

In a gate-field, a node capacitance is connected alternately with a gate, such that one node charge affects the downstream node charge, in a circuit structure such as that shown in Fig.2.17.1. The chain of capacitors connected by the gates stores information during signal processing very much like an RC chain. Yet the information stored in an N-stage gate chain is much larger in volume than that stored in a passive RC chain, and information capacity is proportional to N. This difference originates from the essential difference of the RC chain and the gate chain that the latter includes gain, nonlinearity, and directionality. If unity gain amplifiers are cascaded for many stages, the signal propagates without attenuation, and the information-storage capacity proportional to the chain length is available. If the gain of each amplifier is higher than unity, the gain accumulates over the stages, and the later amplifier stages saturate. In this condition one period of spatial oscillation stores one bit of information. If the gain is less than unity, the signal decays as it propagates, and the information-storage capacity decreases because only so many stages from the driving end are able to store information. Figure 3.05.3 shows the voltage profile of a fifty-stage inverter chain, observed at the even-numbered node index locations. The parameter values used in the numerical analysis is shown in the figure. The period of the sinusoid signal is changed from 5 [Fig.3.05.3(a)] to 40 [Fig.3.05.3(e)].

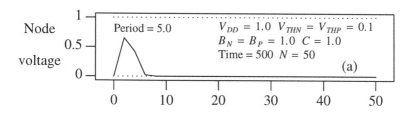

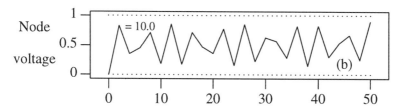

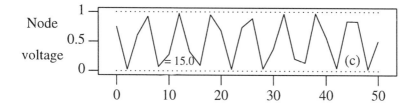

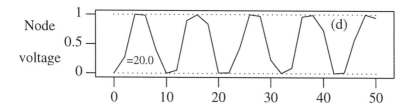

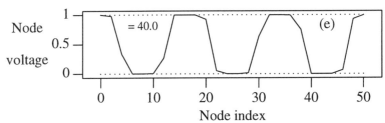

Figure 3.05.3 Information storage
in an inverter field

For a small period the signal does not propagate at all (period 5). At intermediate periods (10 and 15), the voltage profile consists of connected spikes, and the profile is quite irregular, reflecting interaction between the peaks and the valleys of the periodic wave propagating through the chain. At long periods (20 and 40) the spikes are isolated, and there is a flat-top, flat-bottom region between them. In the flat-top or flat-bottom region the inverters are saturated. A saturated inverter prevents transmission of a small-amplitude signal from the input to the output, thereby isolating one spike from the other in the spatial profile. This is a significant information-protection mechanism of a nonlinear circuit, including digital circuit. From the observation, about five stages of inverters are required to store 1 bit of information securely in the chain.

Our next objective is to determine how much information can be packed in the field. The information is represented by the analog-node voltage, which is represented by a cosine wave. Node n voltage is given by

$$V(n,\alpha) = (V_{DD}/2)[1+\cos(\alpha n)] \tag{3.05.1}$$

where the range of $V(n,\alpha)$ is limited from 0 to V_{DD}, and α is a parameter related to the wavelength λ by $\lambda = (2\pi/\alpha)$. Wavelength λ increases as α decreases. If the gate chain is infinitely long, a wave having an arbitrary long wavelength can be accommodated in it. The lower limit of α is then zero. How about the upper limit? In Eq.(3.05.1), n is an integer. If α is replaced by $\alpha+2\pi$, the cosine function gives the same value. $V(n,\alpha)$ is obviously a periodic function of α, with a period 2π. Then if the range $0 \le \alpha \le 2\pi$ is considered, the rest are its repetition. If α is replaced by $2\pi-\alpha$, we obtain the same waveform as Eq.(3.05.1). The maximum of parameter α is then π. At this $\alpha = \pi$, a pair of consecutive nodes at HIGH and LOW logic levels makes a period of the wave. Figure 3.05.4 shows three examples demonstrating how the cosine formula generates the node voltages at the lattice points. Locations of the lattice points are shown on the horizontal axis, by short vertical notches placed at the integer coordinate locations. Figures 3.05.4(a),(b) and (c) are for $\alpha = \pi/6$, $= \pi$, and $= 2\pi-(\pi/6)$. If $\alpha > \pi$, the cosine wave oscillates between two consecutive lattice points, but the values at the lattice points

Macrodynamics of Gate Field

are the only observable values and are the same as those of $(2\pi - \alpha)$.

We assumed that length of the gate-field is infinite. If the gate-field is long but finite, any excitation having the smaller scale than the length can be expressed as a sum of the component waves of the forms

$$V_C(n,\alpha) = (V_{DD}/2)[1+\cos(\alpha n)] \qquad (3.05.2)$$
$$V_S(n,\alpha) = (V_{DD}/2)[1+\sin(\alpha n)]$$

Let the length of the gate-field be an integer N, and the index is set in the range $-(N/2) \le n \le (N/2)$. If an isolated pulse propagates in the chain, a periodic boundary condition at the two ends of the gate-field

$$V(-N/2,\alpha) = V(N/2,\alpha)$$

is set to Eq.(3.05.2). We have $\alpha N = 2\pi m$, or

$$\alpha_m = 2\pi m/N$$

where m is a positive integer in the range from 0 to N. At $m = N/2$, $\alpha_m = \pi$. From the waveforms of Fig.3.05.4 all the cosine waves having wavelengths longer than 2 are included within the range $0 \le m \le N/2$.

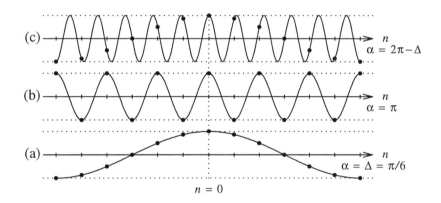

Figure 3.05.4 Traveling wavefronts in a discrete field

Similarly, all the sine waves are included in the same range as well. Then any excitation is written as

$$V(n) = \sum_{m=1}^{[N/2]} [A_{Cm} V_C(n,\alpha_m) + A_{Sm} V_S(n,\alpha_m)] \qquad (3.05.3)$$

where the sum is over all the integers from 1 to $[N/2]$. In this sum, N unknowns A_{Cm} and A_{Sm} are used to match the N node voltages of the gate-field to represent the wave. This is possible, and the result is definite and unique, since the coefficients are the solutions of a set of N-th rank simultaneous equations. This is the most general expression of wave that exists in a discrete chain, not necessarily an inverter chain, having N discrete and accessible nodes. There are N independent sine and cosine components. This is the maximum information that can be packed in the chain.

3.06 Information Stored in a Ring

If we review the objects to which quantum mechanics is applied, we find that there are two types. The first type objects are the quantum-mechanical attributes of a classical particle, such as spin, charge, and mass. The objects do not have a clear and intuitive model of their internal structure: why such an object is observed as such is never convincingly explained by classical analogy, except possibly, spin (Section 2.21).

The other type includes an electron trapped in a central field. The electron state is described by the Schroedinger wave function, and the electron wave demonstrates the characteristic interference. We are able to understand the internal structure as a classical wave. If the models of the two types of objects are sought in digital circuits, the first type is obviously a static latch that retains the state represented by the DC level or by the energy stored in the node capacitance. Its *internal* structure cannot be modeled since the positive feedback that retains the latch states has no deeper-level model. The second type finds its counterpart in various ringoscillator circuits. The wavefront that travels in the structure provides a classical model that has a close parallel to the electron wave in the atom. The two types cover the objects that exist in the quantum-mechanical world. In the gate-field model, it is possible to bring the two into perspective, and the two can be brought into perspective at the fundamental level. To gain more insight we need to study the behavior of ringoscillators. To understand the two types, we need to study a group of circuits that belong to ringoscillators in a quite general definition. To begin with, we study ringoscillators built from inverting and noninverting buffers.

A ringoscillator is a closed-signal path that applies frequency-selective positive feedback. This is a definition based on linear-circuit theory. Whether or not the closed loop breaks into oscillation has been determined using linearized theory by observing the stability of the DC bias point (Shoji, 1992, 1987). I point out that the definition is too restrictive. The small-signal stability study at the DC bias point assumes that all the gates of the ringoscillator interact as the oscillation builds up. This assumption is not necessary. If we use the restrictive definition, only the circuit of Fig.3.06.1(b) having an odd number of inversions of the node transition polarity qualifies as a ringoscillator. The circuit shown in Fig.3.06.1(a) becomes a latch having two stable states. The boundary between a latch and a ringoscillator is, however, never that clear. We are able to show that the definitions of a latch and of a ringoscillator overlap, and this is the crucial idea of the following discussion.

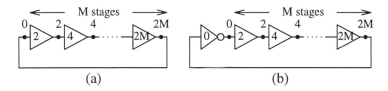

Figure 3.06.1 Buffer closed paths that hold excitation

We assume that an inverting buffer is a conventional CMOS inverter, loaded by capacitance C at the output. A noninverting buffer is built by cascading the two inverting buffers, as is discussed in Section 1.02. We assume this configuration in the following examples of the numerical analysis. If M-stages of noninverting buffers are cascaded to make a closed loop, $2M$ inverting buffers exist in the loop. It is convenient to group two consecutive inverting buffers to a single noninverting buffer. Then a closed loop of buffer is either of the structure of Fig.3.06.1(a) (even number of inverting buffers, $2M$) or Fig.3.06.1(b) (an odd number of inverting buffers, $2M+1$), where M is a positive integer: the former has been considered as a latch, and the latter is a ringoscillator. In the following study, all the nodes, including the unaccessible nodes of the noninverting buffer, are capacitively loaded. The node indices are given to all the nodes, including the unaccessible internal nodes of the noninverting buffers, but we observe the noninverting buffer outputs only, for the convenience of graphic presentation, to avoid cluttering of the figures of the node-voltage profile.

The circuit of Fig.3.06.1(a) appears to settle at either of the two states: all the noninverting nodes (0, 2, 4, ...) have the HIGH logic level or the LOW logic level. They are the only two steady states. They are only absolutely stable states. The closed-loop circuit supports various dynamic activities by setting the initial conditions to the nodes, and this is the crucial additional feature of the general cascaded and closed inverter chain. Practically, this capability is implemented by a switch at each node that short-circuits the node to the Bolean HIGH or LOW voltage levels. At $t = 0$, all the switches are released, and the oscillator goes into activity.

The buffer stages in the quiescent Boolean states that are between two consecutive switching activities work effectively as an isolating barrier between the activity zones. Many switching activities are able to coexist to display a complex pattern, at least for some time. The stability of the activity depends on how many quiescent buffers are between the activity zones. If more buffers in the quiescent state exist between the activity zones, there is more isolation between the activities. Suppose that the length of the loop, $2N$, is large. Choose $2N_1$ and $2N_2$ such that $0 < 2N_1 < 2N_2 < 2N$, and none of $2N_1$, $2N_2 - 2N_1$, or $2N - 2N_2$ is small. Hold the nodes $0, 2, \ldots, 2N_1 - 2$ at LOW logic level, nodes $2N_1, 2N_1 + 2, \ldots, 2N_2 - 2$ at HIGH logic level, and nodes $2N_2, N_2 + 2, \ldots, 2N (= 0)$ at LOW logic level by externally forcing the nodes by the low-impedance outside voltage sources. The buffers $2N_1$ and $2N_2$ have conflicts between their input and output levels, but that never matters, since voltage forcing is active only at $t < 0$. At time $t = 0$, the forcing voltage sources are all disconnected. Then the section having a HIGH logic level moves from the original location $(2N_1, 2N_2 - 2)$ to $(2N_1 + 2, 2N_2), \ldots$, and so on. This mode of operation is shown in the example of Fig.3.06.2.

In Fig.3.06.2, the length of the chain, $2N$, is 30. The even-numbered nodes $0, 2, 4, \ldots, 30 (= 0)$ are the accessible noninverting buffer nodes. Nodes 0, 2, and nodes 12 to 30 (= node 0) were held at the LOW logic level, and nodes 4, 6, 8, and 10 were held at HIGH logic level. At $t = 0$, the forcing voltage sources are all disconnected. In the numerical simulation, the inverting buffers are assumed pullup-pulldown symmetrical, and the FETs are modeled using the conventional low-field, gradual-channel FET model. The parameter values used are

$$B_N = B_P = 1.0 \quad V_{THN} = V_{THP} = 0.1 \quad V_{DD} = 1.0 \quad C = 1.0$$

The section having the high logic level moves at a constant velocity, about 0.5 stages per unit time, around the closed loop. This velocity is consistent with the time constant of the inverter output, estimated at $C/[(B_N + B_P) V_{DD}] = 0.5$. The effective delay per stage is four times that.

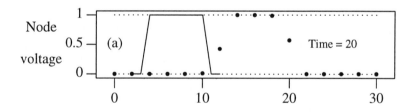

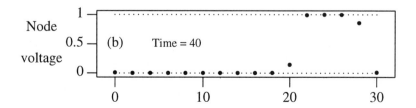

Macrodynamics of Gate Field

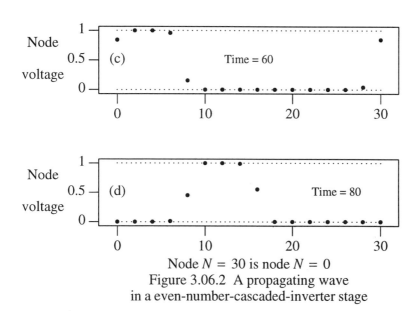

Figure 3.06.2 A propagating wave
in a even-number-cascaded-inverter stage

The motion of the features of the voltage profile is a steady translational motion: time-dependent shape variation of the moving-voltage profile is not clearly observable within the time scale from 0 to 80, from Fig.3.06.2. This means that the switching activities at the pulse's leading and the trailing edges are reasonably well isolated, and they do not interact. Because of this isolation mechanism, the thirty-member loop works in many modes other than absolutely stable latch mode, which has the two stable states.

The initial profile, from which the sequence of Fig.3.06.2 developed, is shown by the solid curve in the first voltage profile. By increasing complexity of the initial profile, varieties of circulating profiles can be established in the loop. Figure 3.06.3 shows a moving profile that has three plateaus, each having widths 6, 6, and 10, in a loop having sixty stages. The three-plateau profile circulates around the loop, maintaining its shape. The profile is essentially stable for a period of time long enough to observe the activity. As we see later, the profile is not absolutely stable, if time scale is much longer that 80.

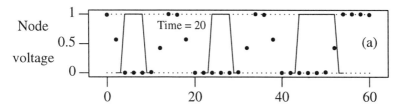

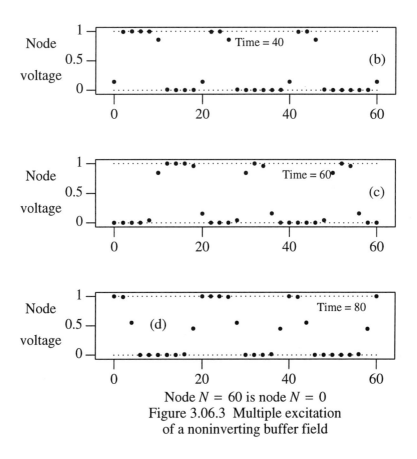

Node $N = 60$ is node $N = 0$
Figure 3.06.3 Multiple excitation
of a noninverting buffer field

A question is, whether the complex waveforms in a noninverting gate loop are really stable over a very long period. Figure 3.06.4 shows a sequence of long-term evolution of an isolated pulse, having the initial profile pulse width spreading over six inverter stages or three noninverting buffer stages. The pulse width gradually decreases over the time interval of $t \approx 2 \times 10^4$ and then decays rapidly at about $t = 26,000$. The pulse-decay mechanism is the same as that discussed in Sections 2.17 and 2.18.

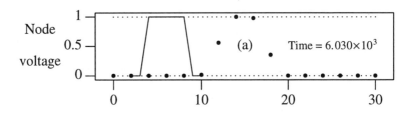

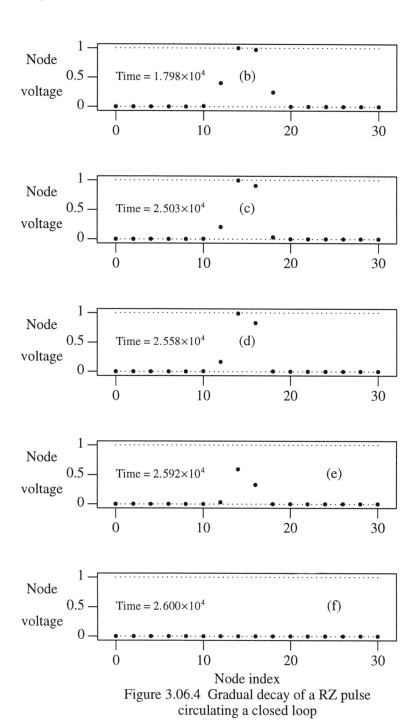

Figure 3.06.4 Gradual decay of a RZ pulse circulating a closed loop

The trailing-edge transition of a narrow pulse starts at the residual voltage of the leading-edge transition, so that the pulse width narrows as it propagates along the cascaded chain. In a closed path the leading-edge transition starts from the residual voltage of the trailing-edge transition as well. This effect is small, however, since the leading edge is behind the trailing edge by a larger number of the buffer stages. The trailing edge affects the leading edge much less than the leading edge affects the trailing edge, in one narrow pulse in a long loop.

Figure 3.06.5 shows the voltage profile of an inverting buffer chain that contains only one transition edge. Since only even-numbered index-node voltages are shown, the end ($N = 37$) node is not shown. This is the node that is identical as node 0. Therefore, two ends have always different logic levels. The profile is absolutely stable. A conventional ringoscillator is operated by this fundamental mode. An inverting-loop ringoscillator is able to support modes other than this, however.

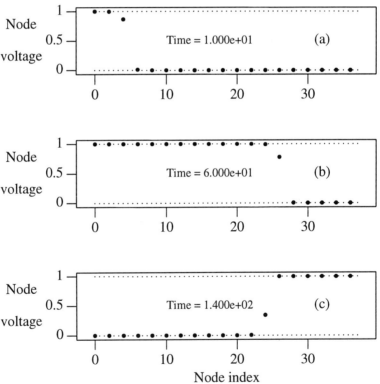

Figure 3.06.5 Ring oscillator in the fundamental mode

Figure 3.06.6 profiles have three transitions: one is at the end shown by the arrow in Fig.3.06.6, and the other two are shown by the solid curve in the first profile, showing the initial condition. The profile is stable, as it circulates along the loop, but as for the long terms, the profile decays to the profile of Fig.3.06.5, which contains only one transition. A single

Macrodynamics of Gate Field

transition in a odd-number member loop is unable to decay as the step-function excitation in an infinitely long gate-field.

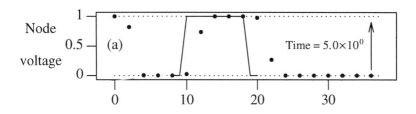

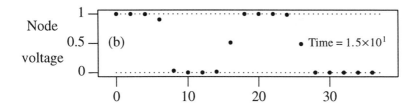

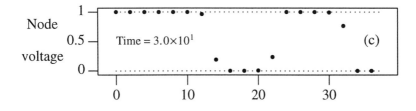

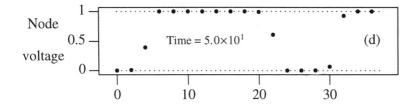

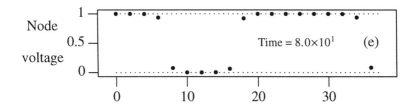

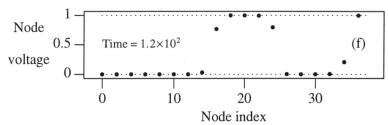

Figure 3.06.6 A ringoscillator excited to a harmonics mode

From the examples of this section, we may conclude the following. Both inverting and noninverting ringoscillator loops sustain various modes of transient activities. In a noninverting gate loop, the fundamental mode is static: it is either all observable nodes HIGH or all nodes LOW. This mode is stable over indefinite time. In this mode of operation, the loop is effectively a binary latch. The *higher* modes of the loop, which include switching activities, decay to either of the two absolutely stable modes after a long time. If the features of the initial voltage profile consist of narrow upgoing spikes, they decay, and the chain settles at the LOW logic level. If the features are narrow downgoing spikes, the chain settles at the HIGH level. This decay mechanism is due to the interaction between the two transition edges of the pattern. In an inverting gate chain, the fundamental mode includes one moving transition in the loop. The mode is absolutely stable, and the loop is kept oscillating forever. Any higher modes decay to this mode after a long time, by the same mechanism as the noninverting gate loop. There is a close parallel to the behavior of isolated pulses discussed in Sections 2.17 and 2.18. The conventional binary latch and ringoscillator are the two essential ground-state operations of noninverting and inverting gate chain loop, respectively. What is important to note is that the oscillator has many higher-harmonic modes of operation, which have various degrees of stability. Some of them last long enough to be useful for signal processing. The loop structures have enormous capability for expressing data and operate on data, and this is the point that has not attracted sufficient attention at this time. The flexibility may have some significant practical consequences.

3.07 Extraction of the Features of Data Pattern

An interesting application of a ringoscillator is to use its capability for accepting an entire complex data pattern, and extracting the dominant features in the time sequence, as the data pattern circulates the ring. The capability of extracting the most important feature from a given spatial figure is not an elementary operation. Figure 3.07.1 shows an example. The solid curve of Fig.3.07.1(a) shows the original data pattern established at $t = 0$. The dominant feature is one RZ pulse, which has a HIGH region wider than the LOW region. The dominant feature has three extra features superposed on it: a is a narrow upgoing pulse, B is a little wider downgoing pulse, and C is a still wider downgoing pulse.

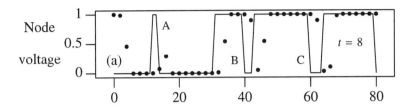

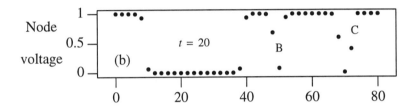

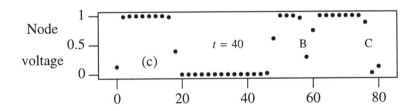

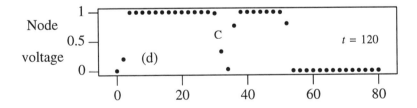

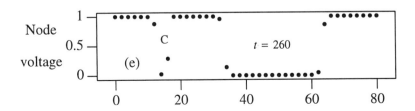

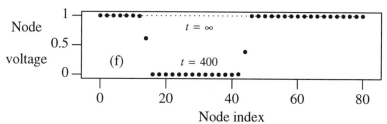

Figure 3.07.1 The filtering effects of the fine features of the pattern on a ringoscillator

A, B, and C may be considered as *noise* or secondary features superposed to the main feature. As the data pattern circulates the ring, first the narrowest pulse A decays and disappear, as observed from Fig.3.07.1(b). Then the second narrow pulse B amplitude decreases as shown in Fig.3.07.1(b) and (c), and it disappears at Fig.3.07.1(d). The widest of the three, pulse C, lasts some more time as shown by Fig.3.07.1(d) and (e), but eventually it deacays, as shown by Fig.3.07.1(f). After a quite long time, even the principal feature, the wide-top RZ pulse, settles at the constant HIGH voltage level. An interesting aspect of this operation is that as the voltage profile circulates, the less important feature is removed earlier than the more important feature, thereby ranking the importance of the features in order. It is possible to judge the relative importance of the features in a complex pattern in sequence. As can be observed from this example, a ringoscillator has a highly intelligent capability for eliminating noise from the data pattern and for displaying the successively more coarse-grained, main features of the data.

3.08 Digital Excitation in a Closed Path

In Sections 3.06 and 3.07 we observed that a closed-signal path of inverters holds a complex data pattern, and if observed at a location on the loop, the pattern appears there sequentially. Although the complex pattern may ultimately decay, it lasts long enough to process the information over many steps, if the pattern satisfies certain robustness criteria. The closed-loop structure has several unique capabilities that may have interesting consequences and, possibly, applications. It is the closest electronic model of the circuit of the central nerve system of an animal's brain. To gain more insight into the problem we need to study the operation as a whole, using a manageably simple and an idealized model. The complex data patterns of the last section are the analog voltage profile within the loop. We adopt a simplification by assuming that the profile is a digital, Boolean-level profile. For this requirement to be satisfied by the electronic hardware, certain restrictions on the structure and the observation time must be set. This is equivalent to building a classical model of a partially quantum-mechanical object. At the beginning of Section 3.06 I pointed out that there are two types of quantum-

Macrodynamics of Gate Field

mechanical objects. Only the second type, which has a classical model, allows this simplification. To make this point clear, let us consider the hardware's structural restriction first.

A closed path is built from a single inverter as the basic unit, and all the output nodes of the capacitively loaded cascaded inverters are observable. If a signal front is at the location of the n-th inverter, the input and the output of a single inverter can be both in the state of transition. During the transition the Boolean-levels of the nodes become uncertain. Can this difficulty be minimized or avoided by choosing the time of observation properly?

Figure 3.08.1 shows successive node-voltage waveforms of a symmetrical inverter chain. As node 1 voltage V_1 goes up, node 2 voltage V_2 goes down, and their transition regions shown by the thick horizontal lines crossing the node waveforms, overlap. They are the regions where the node voltages are in the range between V_{THN} and $V_{DD} - V_{THP}$, and therefore the Boolean-levels cannot be determined with certainty by the quantum-mechanical test equipment (Section 2.09).

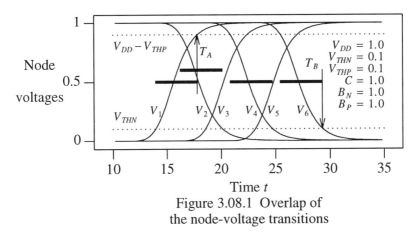

Figure 3.08.1 Overlap of the node-voltage transitions

The transition regions of V_4 or V_6, however, do not overlap with that of V_1. Then, if we use a five-stage cascaded inverter as the basic unit of building an inverting buffer field instead of a single inverter, and if we choose the times of the node-voltage measurements to the right side of T_A and T_B for all the gates in the field, all the nodes of the gate-field will have definite Boolean-levels. What is important here is that there exists a sequence of time points where all the measurable nodes of the gate-field have definite Boolean-levels, if all the building block of the field are identical. In the five-stage cascaded inverter chain in the black box of Fig.3.08.2, the voltages of nodes 0 and 1, which are developed across the delay capacitance C_0 and C_1, are measurable. The voltages of the nodes inside the black box, A to D, are not measurable even if they are developed across the delay capacitance C_A to C_D, since the structure inside the box is defined as not accessible by the model. Capacitors C_A to C_D of the internal nodes store, as we saw in Section 3.05, the data assigned to one period.

The second restriction of the inverter field model is to assume that capacitors C_0 and C_1 are large enough to guarantee that the perturbation by the test equipment to the inverter field is negligible. The measurement should not alter the delay or the measurement time sequence over a long time.

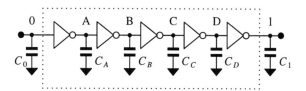

Figure 3.08.2 A substitution to a unit inverter for Boolean-level idealization

To measure the node voltage by the method described in Section 2.09 and to obtain a definite Boolean-level, the time of the measurement must be specified, as we discussed before. This is the peculiarity of building a classical model of a semi-quantum-mechanical object. The method of using the cascaded inverters may be practically difficult to maintain the delays of all the building blocks the same. To do so most reliably and conveniently, a systemwide clock can be introduced. A system that is operated by a systemwide clock is called a *synchronous logic system*. The fundamental building block of a synchronous logic system is a master-slave latch, shown in Fig.3.08.3. The clock CK and its complementary, $\overline{CK}$, cross at the switching threshold voltage of the tristatable inverters, and that is the time when the latch opens to the input signal or closes to it. The clock changes level twice a cycle. In the first change the input level is locked in the master part of the latch, and in the second change the input level is locked in the slave part of the latch. To obtain a precise time specification of the Boolean-level by this method, the data path consists of five stages of cascaded inverting gates, the same as Fig.3.08.2. The input node 0 and the output node 1 voltages are observable, but the internal nodes A through D voltages are not, since they are within the black box, and they cannot be accessed from the outside, by definition. C_0 and C_1 are also large, to make the timing jitter negligible, but the precision requirement is less stringent because of the clocked operation. The examples of the following sections are assumed to be built from the robust, clocked circuit, since we study arbitrarily long-term evolution of the Boolean-level profile of a ringoscillator.

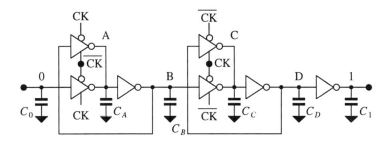

Figure 3.08.3 A clocked inverter for a gate-field

3.09 Multiple Ringoscillators

A ringoscillator consists of a closed loop of inverting gates or building blocks. The loop has an even or odd number of cascaded inverters, and both types support excitation(s) traveling along the loop. If a ringoscillator consists of multiply connected loops, the oscillator will show more complex behavior than just supporting permanent or temporary circulating excitation pattern. To study this problem in various cases, we need a simple but essential physical model of complex inverting gate loops, and a convenient representation of the oscillator states. The simplest graphical oscillator representation is as follows.

A ringoscillator is shown by a closed curve making a rectangle, or connected rectangles in general, that has the same topological connectivity as the oscillator. The nodes are represented by points that are threaded by the curve. A single segment of the curve between two consecutive points represents an inverting gate or a more complex building block. To guarantee that the node's Boolean-levels are definite, two or more odd-numbered stages cascaded inverters, or clocked latch arrangement discussed in Section 3.08 are required as the unit to build the loop. The direction of excitation propagation is indicated by the arrows. The digital level of a node is represented either by a closed circle for the HIGH or by an open circle for the LOW Boolean logic level. An odd- and an even-numbered member ringoscillator is shown by Figs.3.09.1(a) and (b), respectively.

In a steady state the input node and the output node of an inverting gate take complementary Boolean-levels. Then closed and open circles alternate, as is shown in Fig.3.09.1(c). In this oscillator, it stays in one of the two stable states forever. Two stable states exist, since open and closed circles of Fig.3.09.1(c) can be exchanged to get the other stable state. In an oscillating state, the alternating rule must be violated somewhere on the loop. In an odd-numbered member inverting-gate loop, the rule is violated an odd number of times, and the minimum is once as shown in Fig.3.09.1(a). In an even-numbered member inverter loop the rule is violated an even number of times and the second minimum is twice, as shown in Figs.3.09.1(b).

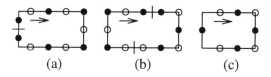

(a)   (b)   (c)

Figure 3.09.1 Ringoscillator representation

The location of the violation is shown by the crossing lines. The location of the rule violation moves along the loop, and the motion is observed as the electrical activity of the loop. What we call an *excitation* is a location where the alternating rule is violated. It is wrong to conclude that an even-numbered inverter loop is simply equivalent to a latch, which is able to represent only two permanently stable states. The normal latch operation occurs only if the length of the loop is short. The loop is able to accommodate varieties of dynamic activity. A huge closed loop is just a latch if the circuit is to provide a pair of static logic levels. If the loop is operated dynamically, however, it is able to represent a complex sequence of HIGH-LOW *patterns* circulating around the loop, by creating a large number of violations to the alternating level rule. In the circuit the variety originates from the spatial frequency spectrum (Section 3.05). Operation as a latch corresponds to the lowest spatial frequency. The loop supports a higher spatial frequency pattern than that.

It is interesting to note that there are significant parallels between the multiple ringoscillator circuit and the neuronic logic system. In the neuronic system the capability of sustaining two steady-state logic levels is not even provided. A single neuron is functionally equivalent to a monostable trigger circuit. Lack of static representation in the evolutionarily perfected system indicates where the signal processing power of a neuronic system resides. The superior capability of a neuronic logic system in displaying and maintaining excitation patterns on the connected neurons, and the complex excitation pattern on a multiply-connected ringoscillator are both originating from the same architectural principle accommodation of a wide range of spatial frequency in a simplified extended-loop structure.

To construct a connected loop, a gate having more than one input must be set at the point where two or more paths meet. The gate can be a NAND, NOR, or any other multiinput gate. At the point where a signal path splits, a single unit drives more than one destination, and this creates no structural irregularity. The two types of joints are clearly distinguished. The point where a signal path splits is a featureless point, and

Macrodynamics of Gate Field

therefore no special indication is required. The point where two paths merge is special, in that a multiinput gate must be placed there. The multiinput gate is represented by the two input lines connected to the confluent point, or the node. The point is enclosed by a rectangle, as shown in Fig.3.09.2(a). The directions of the signal propagation are indicated by the arrows. Not all the arrows are necessary, since often by observation of the path structure the direction can be determined. If a single arrow is placed at the output of the multiinput gate as shown in Figs.3.09.2(b) and (c), the propagation directions can be determined uniquely. In Fig.3.09.2(b) the excitations move from left to right. The point where two or more paths meet is a special point that cannot be moved. The point where a single path splits into two or more can be moved by simply duplicating some units. In Fig.3.09.2(c) an unit B' was added to move the location of the path split from B of Fig.3.09.2(b) to A of Fig.3.09.2(c).

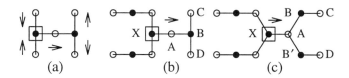

Figure 3.09.2 Excitation-path merge and separation

To study the electrical activities of a complex ringoscillator, it is convenient to use a computer program that displays the development of the loop's Boolean-level profiles. To gain an insight into the loop's working mechanism quickly, we use the Boolean-level analysis methods, assuming the circuit simplification of the last section. The state of a node is characterized by a single Boolean variable that takes the HIGH or the LOW level instead of an analog voltage. We define an integer node index N for each loop that includes a merge or a split of the paths. N starts at the confluent point (or merge) and increases in the direction of the signal propagation. A section of line connecting two consecutive Boolean-level indicators at the nodes contains one inverting building block only. The Boolean variable assigned to node N is written as $B(N,t)$, where t is time. In this model, time progresses discretely, as is discussed in Section 3.08. The state of the loop is defined at the integer time points, $0, \Delta t, 2\Delta t, \ldots, m\Delta t, \ldots$, where Δt is the fixed step. This is the principal simplification of the circuit, and a few more clarifications must be made.

The need for discretizing time is consistent with the discretization of the node voltages to the Boolean-levels as we observed in Section 3.08. Where is the time required for the node voltage change allocated? During the two consecutive time points, the state change occurs. The state of the circuit between the two consecutive time points is excluded as undefinable, since if a measurement is carried out, the test setup may present probabilistic results. By the end of the period a clearly defined state having definite Boolean-levels emerges. Time is not a continuous variable, and neither is the Boolean-level. The time step required between the two consecutive Boolean states of the circuit is the delay time of the gate or the unit building block. This time step must be the same for all the gates and building blocks if no clocking arrangement is used. If a multiinput gate is included in the loop, its delay is generally longer than the delay of an inverter. In our simplified model the multiinput gate is designed so that the delay is same as the rest of the inverters. The case where this assumption is not allowed is discussed briefly later.

If a set of Boolean-levels at time t is given, the set of the levels at $t+\Delta t$ is determined by

$$B(N,t+\Delta t) = B(N,t) \quad \text{if} \quad B(N,t) = \overline{B(N-1,t)}$$
$$B(N,t+\Delta t) = \overline{B(N,t)} \quad \text{if} \quad B(N,t) = B(N-1,t)$$

This algorithm holds in the inverter chain. The first equation is a representation of the alternate logic-level rule that maintains the steady Boolean-level profile if the rule is obeyed. The second is the mechanism to move the point where the rule is violated forward, with time. By choosing the node-index origin properly, a multiinput gate such as NAND2 gate can be placed at the joint between the beginning (node 1 as the output of the combining gate) and end (node N_A and N_B for loop A and loop B, respectively, as the two inputs). Then its output node Boolean-level is given by

$$B_A(1,t+\Delta t) = B_B(1,t+\Delta t) = \text{LOW}$$
$$\text{if} \quad B_A(N_A,t) = B_B(N_B,t) = \text{HIGH}$$
$$B_A(1,t+\Delta t) = B_B(1,t+\Delta t) = \text{HIGH} \quad \text{(otherwise)}$$

If the multiinput gate is a NOR2 gate, the algorithm is modified accordingly.

A computer program accepts a loop configuration specified by several integer parameters. In the program used to produce the following examples, a two-loop ringoscillator configuration was specified by the three integers N_0, N_1, and N_2. N_0 specifies the number of points (or nodes) in the first loop (left side), N_2 specifies the number of points in the second loop (right side), and N_1 specifies the number of points shared by the two loops. The program is required to produce outputs easy for visual inspection. Graphics output capability is included. In the program, the two loops are drawn in a 90 degree rotated Greek character θ shape, the common two loops merge at the top, and the central path splits at the bottom. The direction of the signal in the central vertical path is from top to bottom. The initial state of the ringoscillator (given arbitrary) is shown at the upper left corner of each drawing. The state at $t = \Delta t$ is shown directly below the initial state, and so on. As the bottom of the column is reached, the next is the top of the next column.

The behavior of a multiple-loop ringoscillator is quite complex, but several general conclusions can be drawn by observing the computer analysis results. In the computer analysis, it is necessary to try various initial Boolean profiles of the loop and compare the results. According to the general observation of the trial runs, what matters most in the oscillator behavior is, whether N_0 and N_2 are even or odd numbers. N_1 does not matter, as we observed in relation to the explanation of Fig.3.09.2(b) and (c). N_1 is in a way a spurious parameter, chosen to make neater graphic representation. We chose small N_0, N_1, and N_2 and examine the cases where (1) N_0 and N_2 are both even numbers, (2) one of them is an even and the other an odd number, and (3) they are both odd numbers. We observe that a ringoscillator becomes successively more unstable in the order from (1) to (3), or it is more likely to get in stuck at a steady-state configuration in the reverse order from (3) to (1). A multiple-loop ringoscillator has indeed mixed properties of a latch and an oscillator.

Figures 3.09.3(a) and 3.09.3(b) show that the two possible types of HIGH-LOW alternating voltage profiles established in an even-even loops having parameter values $N_0 = 8$, $N_1 = 3$, and $N_2 = 10$. For simplicity, we write (8,3,10) loop configuration.

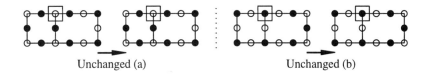

Unchanged (a)　　　　　　　　Unchanged (b)

Figure 3.09.3 Inactive ringoscillators

The Boolean-level configurations are stable. The initial voltage profile does not change with time. The multiply connected loop works effectively as a single binary latch, and the two Boolean states of the latch have the time-independent voltage profiles shown in Figs. 3.09.3(a) and 3.09.3(b). If the HIGH and the LOW logic levels of the points of Fig.3.09.3(a) are exchanged, we get Fig.3.09.3(b). In both cases the HIGH-LOW alternating rule holds. This circuit is quite wasteful as a binary latch, since only two inverters are needed to build it. If building a binary latch were the objective, a coupled even-even loop ringoscillators would make no sense, but the crucial point is that this is not the only operational mode.

A closed even-even number loop is able to support traveling excitation, and this feature adds rich varieties of the data-processing capability of complex ringoscillators. The excitation may not last forever, yet the electrical activity may continue long enough to let the observer recognize its significance and to draw useful conclusions. There are two types of transient and quasi-static phenomenon. The example of the first kind is shown in Fig.3.09.4.

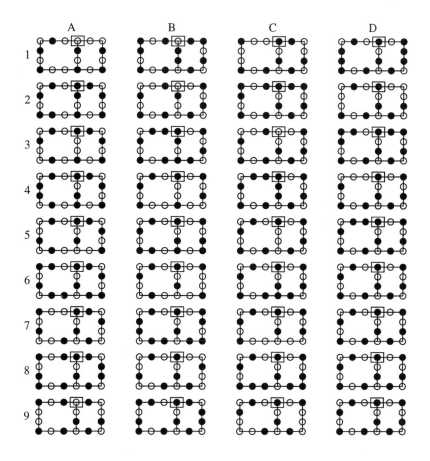

Figure 3.09.4 Transient response of a ringoscillator (even-even)

The figure shows the development of a transient in a (12, 4, 10) coupled ringoscillator, and the initial condition is shown by the voltage profile at the top left corner (A1). The state of the oscillator evolves with time in the sequence (A1), (A2), ..., (A9), (B1), ..., (B9), (C1), ..., (C9), (D1), ..., (D9). The initial profile goes through thirty-two time steps of development and ends up with the final profile shown in (D6). The profile satisfies the alternating HIGH-LOW rule of a stable state. By reaching the stable state the electrical activity is terminated. In the (12, 4, 10) coupled ringoscillator there is the second stable state, obtained by interchanging the open and closed circles of the profile of (D6). Starting from a given initial voltage profile, the ringoscillator ends up with one of the two stable states and settles there. The transient settles absolutely because the circuit arrives at the self-consistent final state satisfying the alternating Boolean-level rule, having no internal conflict. The conflict is created by the initial condition, since the nodes are set at the initial voltage levels and are released at time $t = 0$.

In certain special cases the internally self-consistent final state is never reached. An example is a symmetrical two-loop oscillator (the left and the right loops are identical), having identical initial Boolean-level profile. This oscillator is equivalent to a single-loop oscillator, and this is really a trivial case. Even if the loops have even-numbered members, the oscillation appears to sustain forever, but since the front and rear ends of a pulse interact, the pulse width shrinks, and the excitation dies after some time. This ringoscillator also ends up with the internally self-consistent steady state but by a fundamentally different mechanism. The example of Fig.3.09.4 is by logical consistency, and this case is by the natural pulse decay. The two mechanisms have a qualitatively different time constant to settle at the final state.

A coupled ringoscillator having even-odd or odd-even configuration may get stuck in a steady state, but by a different mechanism from an oscillator having an even-even configuration. In the inset of Fig.3.09.5 (D6) a steady state of a two-loop ringoscillator having an (8,3,7) configuration is shown. The eight-member left-side loop gets in stuck at the state that disables the NAND2 gate at the confluence point permanently. Then the seven-member right-side loop ceases to pass excitation through the confluent point, and therefore the loop is unable to oscillate. The left-side loop obeys the HIGH-LOW alternating rule for stability, but the right-side loop does not.

The same (8,3,7) coupled ringoscillator is able to oscillate continuously, however, if the initial condition is chosen differently. Starting from the initial condition (A1) of Fig.3.09.5, the oscillator voltage profile develops seven time steps to profile (B2), and from the profile the oscillator goes into a periodic operation. The same state reappears eight steps later at profile (C4). This example has a similar feature as Fig.3.09.4(a), which goes through thirty-two steps to reach the steady state. As for state development, whether a ringoscillator ends up with a steady state or a steady oscillating state is evolutionally same. It would be surprising if the number of the members of a loop determines final states which are qualitatively different. The steady state and steady oscillating state indicate absence of localized conflict in the ringoscillator. In the fundamental oscillation mode of ringoscillator, there is a single Boolean-level conflict, but it moves one location to the next uniformly, and the conflict is smeared out over the entire loop. Then the ringoscillator as a whole is in a state that has no conflict. The initial adjustment phase that brings a ringoscillator to the final steady or steady-oscillating state is for removing conflict which the initial state had. This is a basic feature of a ringoscillator-based logic operation. The ringoscillator circuit tries to remove the conflict of the initial state and tries to reach a consistent final state.

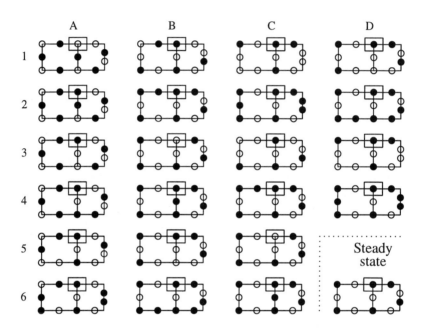

Figure 3.09.5 Active ringoscillator configuration (oscillation)

From stability viewpoint even- and odd-numbered member loops have a fundamental difference: that the former has a stable state that it can be forced into or can be migrated into, and the latter has an equivalent stable oscillating state, both playing the same evolutionary role. From this viewpoint a two-loop coupled ringoscillator having an odd-odd configuration should always oscillate. This point can be made plausible in two different ways. First, if an odd-odd ringoscillator had a steady state, there must be a conflict in each loop, or at the gate of the confluent point. A conflict in a loop moves, so the state cannot be steady. A conflict at the confluent point is self-destructive, so the state cannot be steady also. Second, in an odd-odd ringoscillator it is impossible for one loop establishing an alternate HIGH and LOW Boolean-level pattern that disables the gate at the confluent point. This statement can be rephrased

Macrodynamics of Gate Field   231

as follows. An even-numbered member loop can exercise complete control over the other loop some of the time, but an odd-numbered member loop cannot. The control by the even-number member loop is not complete, since as we see in the later part of this section, a gear mode of oscillation exists. Absence of steady state in an odd-odd ringoscillator and existence of the gear mode suggest that even- and odd-numbered member oscillators are not perfectly symmetrical. In the (9, 3, 7) coupled ringoscillator of Fig.3.09.6, the initial state, (A1), reappears sixteen steps later at (C5), and the cycle repeats.

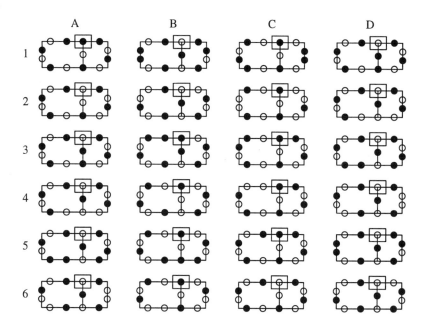

Figure 3.09.6 Active ringoscillator (odd-odd)

Naturally, the example for the reversed initial profile of Fig.3.09.7 oscillates with the same period, but the initial state does not reappear during the later development. Instead, after two steps of development to (A3), the state reappears sixteen time steps later at (C1). The periodic oscillation is to go through a sequence of selected states again and again. If the initial state happened to be one of the selected states the oscillation continues regularly. If not, the initial state evolves several steps to reach one of the selected states, and then the oscillation becomes periodic. The complex dynamics of a ringoscillator can be understood from the state space representation of a ringoscillator discussed in Section 3.12.

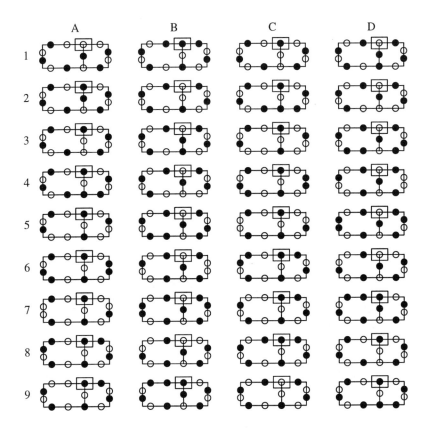

Figure 3.09.7 Active ringoscillator (odd-odd) II

This is the same observation as before: if the initial state is not a correct state to be included in a periodic oscillation sequence, a few steps of correction take place first, and then the oscillator goes into one of the states belonging to periodic oscillation, and from there a steadily oscillating state begins.

The period of oscillation of an odd-odd ringoscillator is derived from the computer analysis based on the simple model. Figure 3.09.8 shows the period measured by the time steps required to return the starting state, versus N_0 and N_2. N_1 was chosen either 3 or 4, and it does not matter. The period is given by

$$\text{Period} = N_0 + N_2$$

This formula was derived by observing many examples and is not a mathematically proven formula. Yet the formula is quite convincing if we think as follows. A complete cycle of oscillation consists of two parts. In the first part, the excitation circulates the first loop once, and in the second part it circulates the second loop once. After completing a tour of the θ-shaped path, the same state reappears. Let us observe how this mechanism works in some detail by an example.

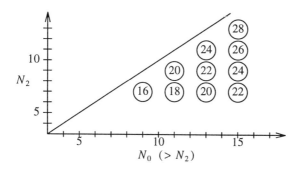

Figure 3.09.8 Period of cyclic oscillation

In Fig.3.09.7 (B2) an excitation front is launched by the NAND gate at the confluent point, which makes a HIGH to LOW transition. Both of the two inputs of the NAND gate became HIGH one time step before, at (B1). The inputs stay in the HIGH level until the newly launched front travels the shorter, seven-member loop and comes back to the input of the NAND gate as the low level at (B8), seven time steps later than (B1). At this point the NAND gate is enabled by the input from the larger nine-member loop (the left side input of the NAND) because the input node cannot change state until the front travels through the longer loop. The LOW level from the right side input passes the NAND gate at (B9) and launches the second LOW to HIGH front at that time. The first front in the longer, nine-member loop arrives as the LOW level later at (C1), but at that time the NAND gate is already disabled by the input from the right loop, and the state of the gate does not change. The second front propagates the shorter, seven-member loop and arrives at the input of the NAND gate at (C6), but it is unable to pass the gate since it is disabled by the LOW input from the left loop. The second front travels through the left loop and returns to the input of the NAND gate as the HIGH level at (C8), and the same state as at (B1) is reproduced. In this mechanism the excitation indeed goes through the two loops and returns the starting point after $N_1 + N_2$ steps. This is certainly a nontrivial mode of oscillation of a multiple-loop ringoscillator.

In a multiple-loop ringoscillator, the data patterns of two or more loops interact, and the oscillator modifies, creates and displays complex time-dependent data patterns. The data patterns vary. It is interesting to ask whether there is any way to maintain the same data pattern all the time. If the numbers of the members of a two-loop ringoscillator have an integer ratio, the ringoscillator has such a mode of operation, which is quite similar to the operation of a mechanical reduction gear.

Figure 3.09.9 shows a pair of engaged gears. If the left-side gear rotates eight times clockwise, the right-side gear rotates once counter-clockwise. The number of rotations is inversely proportional to the number of the teeth.

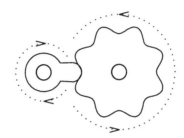

Figure 3.09.9 A gear model of a multiple-loop ringoscillator

This mechanism finds an electronic analog model in a two-loop ringoscillator. Each gear is identified to a loop. The engaged section of the gear is identified as the common path of excitation and the confluent point of the two paths. The gear teeth are the location of the moving Boolean profile in the loops. If one loop has twice more gates than the other, and if the initial condition is set such that the moving Boolean patterns of the two loops mesh at the confluence point once in a cycle of the larger loop and twice in a cycle of the smaller loop, the gate at the confluent point receives the same signal at the two inputs, and the multiple ringoscillator works effectively like a 1 to 2 reduction gear. Figure 3.09.10 shows an example. In this example the left loop has eight members, and the right loop has sixteen members. In Fig.3.09.10, the initial state of (A1) reappears eight time steps later at (B4) and then sixteen steps later at (D2). Eight is the number of gates in the smaller loop. Since the both loops have an even number of gates, the data pattern in the smaller loop repeats every eight steps, and for every two cycles of the smaller loop the larger loop completes a cycle. Gear mode of operation occurs even in the even-even ringoscillator that is more likely to get in stuck at the steady states than to oscillate. It is interesting to note that the operation of eight and sixteen member coupled loops cannot be used to count down the frequency of oscillation of the smaller loop. In a coupled ringoscillator there is nothing to indicate the present state of the larger loop that has by an integer m times more gates. If the Boolean profile passing a location is observed, there are m patterns that are all identical to the pattern in the smaller loop but which m pattern arrived cannot be

Macrodynamics of Gate Field

determined.

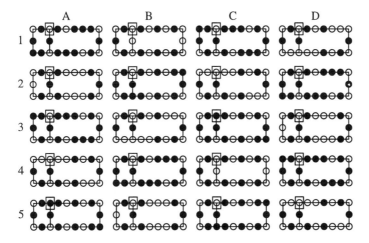

Figure 3.09.10 Gear operation of E-E (8,16) ringoscillators

This is most easily explained by pointing out that the two loops, identified as a pair of meshed gears, do not have a fixed *shaft*, by which the frequency of rotation of each gear can be taken out as information. A shaft fixed to a gear is able to sense the state of all the teeth together, and it is able to indicate the state of rotation of each gear. By observing only the identical teeth of a meshed gear it is impossible to determine the rate of rotation at all. A gear is a device whose state cannot be determined by a local observation. In the gear analogy, a two-loop ringoscillator is like a gear having no shaft fixed to it or a gear that fit into a shaft fixed to the frame (such as the second gear of the three sets of gears used to reduce the rate of rotation). The setup is rather similar to a conventional rotating printer of newspaper: if the drum rotates once, one page of newspaper is printed but it is impossible to determine the *rate* of *rotation* of the newspaper, since all of them are the same: to do so the total number of copies of the printed newspaper must be counted. This is perhaps one crucial feature of a ringoscillator-based logic system.

3.10 A General Observation of Ringoscillator Dynamics

In Section 3.09 we studied various two-loop ringoscillators by examples, to gain a general idea of the ringoscillator problem. In this section we proceed to study the modes of more complex three-loop ringoscillators, and we attempt to build a more systematic theory than we did in the last section. The study is based on the gedanken experiment, a computer experiment, and a generalization.

The structure of two-loop ringoscillators is defined from the connectivity, Greek character θ. The directions of excitation propagation in the sections are determined uniquely from it. If the ringoscillator is more complex than that, the connectivity alone does not determine the direction of the excitation propagation. Figure 3.10.1(a) shows a connectivity of a three-loop multiloop ringoscillator. To find all the possible structures having this connectivity, the following procedure is used. Assign two possible propagation directions to each of the six sections, and generate a total of $2^6 = 64$ possible structures. Of the sixty-four structures, any structure having a node where all the excitations are incoming (converging node) or all the excitations are outgoing (diverging node) is removed. A total of twenty-six structures are left. Some of the structures are duplicates of others, by symmetry or by exchange of the path identification. The duplications are removed as well. Then we obtain the four essential structures shown in Figs.3.10.1(b)-(e). Although structure (e) is consistent with the criteria, the two circled gates must have two outputs, so (e) is not considered.

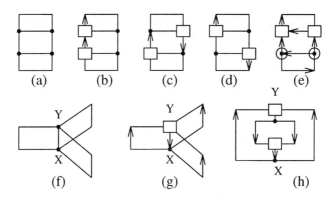

Figure 3.10.1 Connectivity varieties of multiple-loop ringoscillators

In a circuit connectivity shown in Fig.3.10.1(f), there are two distinct structures. In the one shown in Fig.3.10.1(g) there are three sections from nodes X to Y having upgoing propagation direction, and in the one shown in Fig.3.10.1(h) there are two upgoing and two downdoing sections. A three-input gate combines the signals at node Y in the first structure, and two two-input gates combine the signals at X and Y in the latter. The two examples show that, in general, the directions of signal propagation must be shown to specify the structure of a ringoscillator, in addition to the connectivity.

For the convenience of the ringoscillator study, we restate the rule

Macrodynamics of Gate Field

of the Boolean-level profile development in a ringoscillator, focusing the evolution not at each location of the propagation path but rather at a long section of it. This is especially convenient in an inverting multiple-loop ringoscillator, since inversion creates complexity in thinking. We consider section 1 to 6 of Fig.3.10.2(a). If the Boolean-levels of this section are known at time point t, the Boolean-levels of section 2 to 6 are determined at the next time point $t+\Delta t$. Node 1 is driven by a combinational gate, where a different rule applies. In general, the profile at $t+\Delta t$ is determined as follows. First, move the profile at t shown by α of Fig.3.10.2(e) by one step to the right as shown in Fig.3.10.2(e) β, and then take the Boolean complement, as shown in Fig.3.10.2(e) γ. The profile γ is the Boolean profile at time point $t+\Delta t$. We give examples of several patterns on the section, which occurs frequently. Three examples are shown in Figs.3.10.2(b)-(d).

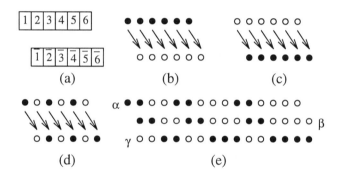

Figure 3.10.2 Development of section Boolean profile

The simplified model of a ringoscillator has a mode that is common to all the structures. If the oscillator is started from the state in which all the nodes are at the Boolean LOW level, the oscillator goes into the state in which all the nodes are the HIGH and then comes back to the initial state in which all the nodes are the LOW. In this mode all the nodes go up and down together as shown in Fig.3.10.3. We call this a *synchronous oscillation mode*. At the confluent point of the paths, the NAND or NOR gate works effectively as a degenerate inverter, since the multiple inputs have always the same Boolean-level.

A multiple ringoscillator has steady states that maintain itself indefinitely, without any change. The studies of the two-loop ringoscillator have shown that an even-even configuration has such states but an odd-odd configuration does not. Let us investigate the steady states of the three-loop oscillator, having the structure shown in Fig.3.10.1(d). The three-loop ringoscillators having the structure, and the numbers of the members in the three loops either even or odd, are covered in the following study. Figure 3.10.4 shows the result. The direction of propagation is shown by the arrows at the output of the multiinput gates. Even (E) or odd (O) numbers of the members in the loops are indicated along with the steady-state Boolean-level profiles. No steady state was found if all three loops have odd-number members.

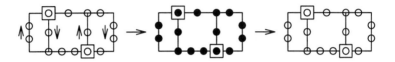

Figure 3.10.3 The synchronous mode of a ringoscillator

Figure 3.10.4 shows that the E-E-E structure has two stable states. If the Boolean-level profile of one is complemented, the profile of the other is derived. The structure works effectively as a latch that consists of the three latches coupled together.

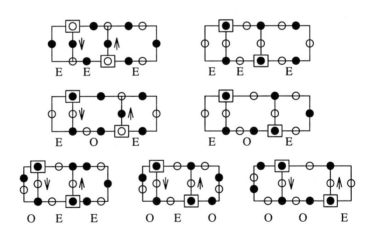

Figure 3.10.4 Steady states of a three-loop ringoscillator

The coupling forces the states of the three latches to create two stable states of the combined-latch circuit. This conclusion holds independent of the number of the members of each of the three loops, as long as the numbers are even. A ringoscillator having E-O-E structure has two stable states, but the Boolean-level profile of the two states are not mutually complementary. The Boolean-level profile of the left loop remains the same. In the central loop there is a violation of the alternating level rule for stability, at the top (left) or the bottom (right) NAND gate. At which of the two gates the violation occurs creates two variations of the steady states. This violation does not matter, since the gate is disabled by one input. The NAND gates output their default HIGH logic level. The three-loop latch degenerated into a single-loop latch by the disabled gate. The two stable states are the states of the right-side loop. We note that the left and right loops are not equivalent in creating the steady states, since there are four inversions from the top left gate to the bottom right gate, but there are only three inversions from the bottom right gate to the top left gate. The three loop oscillators having an O-E-E, O-E-O and O-O-E structure have only one stable state. They are the equivalent to a monostable latch. As we observed, the stability of a ringoscillator generally decreases, as the number of the loops having an odd-numbered member increases.

The synchronous oscillation mode of Fig.3.10.3 and the steady states of Fig.3.10.4 are the features shared by most of the multiple-loop ringoscillators. Those special states are the final states, evolutionally reached starting from many initial states of the oscillator. Figure 3.10.5 shows that a particular initial state of the O-O-O (5-9-7) oscillator evolves after twenty-two time steps to the synchronous oscillation state. Figure 3.10.6 shows the evolution of two particular initial states of the O-O-E (7-9-4) oscillator: the one ends up with the steady state of Fig.3.10.4 and the other with the synchronous oscillation mode. The synchronous oscillation and the steady states are useful in classifying the operation of multiple-loop ringoscillators. How a given initial state evolves into the final state or mode is important in understanding the types of data processing carried out by the ringoscillator-based logic system.

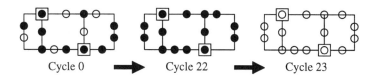

Figure 3.10.5 An O-O-O oscillator settling at the synchronous mode

The synchronous oscillation starts if all the nodes of a ringoscillator are initially HIGH or LOW. The steady-state is reached if the steady-state Boolean profile itself is given as the initial state. An insight into the ringoscillator behavior is gained if we consider the initial states to be only slightly different from the particular initial states mentioned above. The concept of *neighborhood* of a given state is an useful concept. Let the state shown in Fig.3.10.7(a) be given. This is a state of an O-O-O (5-9-7) ringoscillator that has fifteen nodes, and the Boolean-levels of the nodes are shown by the open and closed circles. If the Boolean-level of any one of the fifteen nodes is changed to its complement as shown in Fig.3.10.7 (b) by an asterisk (*), the new state is a neighbor to the original state. The neighbor distance is measured by how many node Boolean states are changed. Figure 3.10.7(c) shows a neighbor to the original state as well, and the distance is 2, since the states of the two nodes indicated by the asterisk have been changed. To gain insight into the ringoscillator, we begin with the first neighbor states to the special initial states.

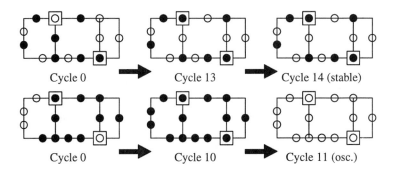

Figure 3.10.6 An O-O-E oscillator settling at the steady state and at the synchronous mode

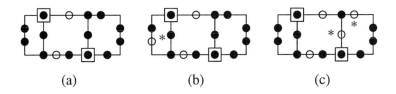

Figure 3.10.7 The neighborhood of a state of a ringoscillator

Table 3.10.1 summarizes oscillator development when the first neighbor states of the steady, and the synchronous oscillation states as the initial states, of the six different structure of the three-loop ringoscillators, having the configuration shown in Fig.3.10.1(d). The number of the nodes in each loop, and the total number of nodes are shown in the third and the fourth rows of the table, respectively. All the structures have about the same number of nodes, ranging from thirteen to fifteen. The E-E-E structure and the E-O-E structure have two stable states shown in Fig.3.10.4 top and the center. The O-E-E, O-E-O, and O-O-E structures have one stable state each. The O-O-O structure has no stable state. The synchronous oscillation is able to start from either the all node

LOW or all node HIGH initial state.

The E-E-E ringoscillator has fourteen nodes, and therefore each stable state has fourteen first-neighbor states. Starting from the fourteen first-neighbor initial states of the first-stable state (top left of Fig.3.10.4), the oscillator is never able to reach the same steady state. The oscillator continues to oscillate with a period of 4, and the mode of oscillation is not the synchronous mode. As a clear contrast to this observation, the oscillator reaches the second steady state (top right of Fig.3.10.4) if it starts from any of the fourteen first-neighbor states of the second steady state. This means that there are significant differences in the *stability* of the steady states against small perturbation. A similar difference is observed on the two stable states of the E-O-E oscillator. The first steady state is shown in Fig.3.10.4 center left, and the second steady state is shown to the right. The single steady state of the O-E-E or O-O-E oscillator has about the same stability, and that of the O-E-O oscillator has a little less. If the oscillators do not end up with the steady states they oscillate, but they never oscillate in the synchronous mode. No case of first-neighbor state of a stable state evolves to the other stable state was detected. The two observations are not general since they are the results of the case studies, but they suggest that *distance* between the two stable states, and between a steady state and the all HIGH or all LOW state, is long. The concept of distance is vague but is useful in understanding oscillator behavior.

Table 3.10.1: 3-loop ringoscillator modes						
Config.	EEE	EOE	OEE	OEO	OOE	OOO
Example	488	496	588	587	794	597
Total	14	13	15	14	14	15
Steady state 1	0	1	12	6	10	*
Steady state 2	14	7	*	*	*	*
Sync. osc. (H)	6	4	5	3	9	15
Sync. osc. (L)	8	8	7	5	10	15

The number of the first neighbors to all HIGH or all LOW states that end up with the synchronous oscillation does not depend so strongly on the structure, but still the structures having more odd-numbered member loops have more first-neighbor states evolving into the synchronous oscillation. In the O-O-O oscillator all the first-neighbor states end up with the synchronous oscillation mode. This observation reminds us that all the first-neighbor states of the second stable state of the E-E-E oscillator ended up with the stable state. The stable state and the synchronous oscillation have the same position in the set of the evolutionary patterns of multiple-loop ringoscillators.

3.11 Modes of Oscillation

In the last section the properties of multiple-loop ringoscillators are studied by focusing attention on their two peculiar and most common operational modes the synchronous oscillation mode and the steady states. A multiple-loop ringoscillator may operate in more than one mode: starting from various initial conditions, the oscillator ends up with varieties of these operational modes. Figure 3.11.1 shows the three distinct modes of oscillation of an E-O-E (4-9-6) ringoscillator. The oscillator has the synchronous mode having period 2, a mode having period 4, and a mode having period 6, in addition to the two steady states.

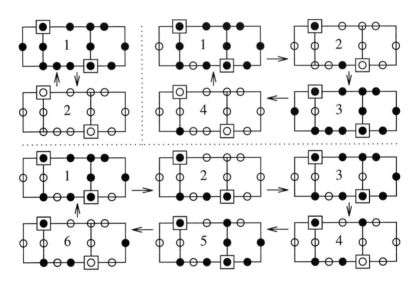

Figure 3.11.1 Various oscillation modes of an E-O-E oscillator

The period-4 and the period-6 mode Boolean profiles of the E-O-E (4-9-6) oscillator show the two mechanisms of multiple oscillation modes. In the period-4 mode, the six-member loop on the right side oscillates in the synchronous mode, having period 2. The right loop includes a NAND gate. For the loop to oscillate in the synchronous mode, the NAND gate input from the left must be HIGH whenever the input from the right loop is HIGH (this is required to generate LOW at the gate's output at the next time point). This condition is satisfied by the mode of the oscillator because the right-loop oscillation is synchronous with the oscillation of the left loop. The NAND gate of the left side loop must satisfy the same condition as that of the right loop. This condition is indeed satisfied again, since the two oscillations are synchronized. Period 4 is the time required for a Boolean pattern to circulate once the four-member loop. In the period-6 mode, the four-member loop on the left side is stuck at the steady state, which enables the NAND gate of the six-member right loop permanently. Period 6 is the time required for a Boolean pattern to circulate the right loop once. By summarizing the two observations, one of the loops of a multiple-loop ringoscillator may go into simple mode of operation, such as a steady state, or a synchronous mode oscillation. If the loop goes into a steady state, the loop's activity ceases. The loop's output enables or disables the multiinput gates of the other loop(s), and the loop having a disabled gate ceases activity. The active region in the oscillator structure is restricted by this mechanism. If a loop goes into the synchronous mode of oscillation, the rest of the oscillator generates the signal that enables the gate. This means that the oscillator activity is synchronous everywhere, and the period is even. A multiple ringoscillator searches for the consistent final state from the given initial state. Consistency seeking is an intelligent capability of the ringoscillator-based logic operation, which may find new applications.

As we observed in the examples, how the gates at the confluent points on the signal paths work is crucial in the oscillator function. Although the logic operation is simple, if the gate controls a data stream and if the Boolean profile in the stream is the issue, it is convenient to represent the gate operation as the algorithm of spatial transformation, as follows. In Fig.3.11.2(a) the square shows a two-input NAND gate. If the control input is HIGH, the upstream data pattern shown by the × marks reproduces faithfully $2n$ time steps later, at $2n$ locations downstream. Here $2n$ is an even integer. At time $2n+1$ the complement sequence of the original pattern appears at $2n+1$ locations downstream. An upstream pattern 110 (in the order seen in the figure) appears as 110 four-steps downstream at $t = 4$, and 001 five-steps downstream at $t = 5$. If the gate is disabled, the NAND gate produces default HIGH logic level. Figure 3.11.2(b) shows the signal propagation path in this case schematically. After $2n$ time steps, up to $2n$ locations to the downstream of the gate are converted to the alternating pattern of HIGH -LOW Boolean-levels starting at the NAND gate location, which is HIGH.

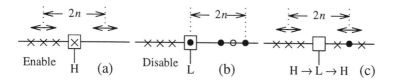

Figure 3.11.2 A disabled gate as the excitation source

If the gate is a NOR gate, the default logic level of a disabled gate is LOW, and the downstream pattern is an alternating LOW-HIGH pattern starting at the gate location, which is LOW. If the NAND gate is disabled at only one time point as shown in Fig.3.11.2(c) the Boolean-level at the location in the data stream is converted to the default HIGH logic level. If the gate is a NOR gate, the Boolean-level is converted to the LOW logic level. The statements made in this paragraph are only a rephrasing of the logic operation, but it is convenient to use this restatement in an algorithm to produce the results mechanically.

An immediate consequence of the spatial algorithm is that it is not necessary to wait the same number of time steps as the number of the members of the loop to reestablish the same Boolean-level profile at a location on the loop. If the same pattern is stored N times in a loop, the loop as a whole operates at $1/N$ times as frequent as the rest of the circuit. The N-times reproduced patterns are all equivalent. The gear operation of multiple loops discussed in the last section uses this mechanism.

A multiple-loop ringoscillator may have more than one mode of oscillation, but there is one or a few dominant periods of oscillation, especially in a simple multiple-loop oscillator. The dominant periods are observed most frequently, as the initial condition of the oscillator is changed in the trial. Dominant periods of the oscillator structure of the last section are summarized in Table 3.11.1. A dominant period may or may not have the repetition of the same Boolean-level profiles in the oscillator loops. By studying the mechanisms of the dominant mode we gain further insight into the oscillation mechanisms.

Table 3.11.1: Dominant periods of oscillation			
Structure	Example	Nodes	Dominant period
E-E-E	4-8-8	14	2,4
E-O-E	4-9-6	13	2,4,6
O-E-E	5-8-8	15	2,8
O-E-O	5-8-7	14	2,8
O-O-E	7-9-4	14	2,4
O-O-O	5-9-7	15	2,14

(1) Gear operation

The period 4 oscillation of the E-E-E (4-8-8) oscillator is a gear-mode oscillation. The best way to understand this mode operation is to remember that a two-input NAND gate is logically equivalent to an inverter if the two input signals are always at the same Boolean-level. In the oscillator structure shown in Fig.3.11.3(a), loops 1 and 3 circulate the Boolean-level patterns, and loop 2 couples the two active loops. The Boolean-level at node B is the complement of the Boolean-level at node C one time step before, and similarly the Boolean-level at node E is the complement of the Boolean-level at node D one time step before. They are both determined by the Boolean patterns circulating in the loops 1 and 3. If the Boolean patterns are chosen such that node A and node B have always the same Boolean-level, and so do nodes E and F, a gear

Macrodynamics of Gate Field 245

operation takes place. If the data pattern in the small loop 1 is repeated N times in the large loop 3 and if the two patterns are properly aligned, an $N : 1$ gear operation takes place.

(2) Enabled loop operation

The period 6 operation of the E-O-E (4-9-6) oscillator is generated by the permanently enabled six-member right-side loop. With reference to Fig.3.11.3(b), the loop is permanently enabled because the four-member left-side loop is stuck at the steady state, in which node D is HIGH level and node A is LOW level. The NAND gate of the left side loop is disabled: the varying node B logic level that is derived from loop 3 activity never matters. The default HIGH level of the loop 1 NAND gate drives node D and G to the HIGH level, which enables the loop 3 NAND gate permanently. A variation of this mechanism is the period 4 operation of the same E-O-E (4-9-6) oscillator. In the mode the six-member right-side loop oscillates in the synchronous mode, as is shown in Fig.3.11.1. A node B level provided by the synchronous mode does not always enable the NAND gate of the four-member left-side loop. When the gate is disabled, however, node A and node B take the same Boolean level in this mode of operation. Then this mode may be considered to be a hybrid of the gear mode and the enabled-loop mode. The period, 4, equals the number of the members of the left-side loop. Steady-state and synchronous oscillation are equivalent because they are both simple, and that makes consistency seeking easy.

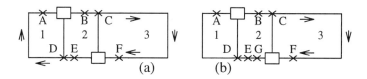

Figure 3.11.3 Understanding the gear
and the enabled-loop operations

(3) Wait and synchronization operation

Let a multiple-loop ringoscillator be made of a small and a large loop, and let the large loop be less than twice the small loop (the size of a loop equals the number of the members). The natural oscillation period of the large loop is longer than that of the small loop. For the combined loop to oscillate at the period of the large loop, the small loop must be able to halt operation at its final state and wait until the next cycle of the large loop begins. Figure 3.11.4 shows such an operation of the O-E-E (5-8-8) oscillator. This oscillation mode, having the dominant period 8, is observed starting from many initial conditions. The NAND gates launch the logic level, which is the complement of the default level at the disabled state, once every cycle. In the left loop this occurrs at time 3, and to the right loop at time 7. The signal propagates along the left loop through times 4, 5, and 6, and at 7 it arrives at the last node of the left loop that is, at the input of the NAND gate of the loop. This takes only five time steps, which is shorter than the period by three steps. During the extra three units of time, the Boolean-level profile in the left loop goes through the readjustment phase to satisfy the alternating-logic level rule as much as possible for stability, and then stays unchanged, waiting for the arrival of the HIGH level from the right loop. The HIGH level arrives at time 2, and the low level is launched at the gate to the right loop at time 3. The period 8 oscillation of the right loop continues uninterruptedly during the time when the left loop is in the wait state. This simple mechanism of synchronization provides a slack to the shorter-period oscillation.

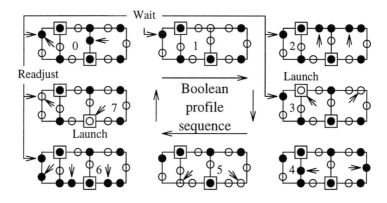

Figure 3.11.4 A wait and synchronize operation

(4) Multiple-loop operation

In modes (1) to (3) the period of oscillation is determined by one of the loops of the oscillator. In the O-O-O (5-9-7) oscillator listed in Table 3.11.1, the dominant period of oscillation, 14, is larger than any periods directly provided by any loop. The period might be interpreted as the period of the seven-member right-side loop, but such an interpretation is unjustifiable in a multiple-loop oscillator operation, since the left- and the right-side loops interact. Figure 3.11.5 shows the evolution of the Boolean-level profile of the oscillation. A close observation reveals the signal flow as shown by the inset of the figure. Nondefault LOW logic level is launched by the NAND gate of the right-side loop at A2 of Fig.3.11.5. The signal circulates the right loop, returns the same gate seven time steps later at B4, and launches the default high level. The signal circulates the right loop, returns seven time steps later at A2 of the next cycle, and launches the nondefault low level again. This is the operation of the seven-member right loop, as a conventional ringoscillator.

This operation is meshed with the operation of the left-side loop. The nondefault LOW level launched to the right-side loop at A2 reaches the NAND gate of the left loop at three steps later at A5 and launches the nondefault LOW level at B1. The signal circulates the left loop, returns the same gate five time steps later at C1, and launches the default HIGH level. The signal returns the NAND gate of the right loop four time steps later at A1 and launches the nondefault LOW level at A2. This second process has period 14, the same period as the first process. This is the mechanism of oscillation. Matching of the period of many processes is required to establish a periodic operation of a multiple-loop ringoscillator. This is a highly intelligent capability of seeking for consistency, a unique feature of the ringoscillator-based logic.

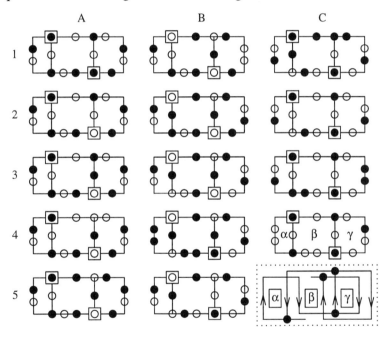

Figure 3.11.5 Multiloop signal propagation
in an O-O-O ringoscillator

The modes of the typical small-scale multiple-loop ringoscillators are discussed in this section. This list gives a more general idea about the oscillator than the examples of the previous sections, but the list is in no way complete. Adding more mechanisms to the list will eventually bring the properties of a multiple-loop ringoscillator-based logic system into perspective. The logic system is quite unconventional, and it is hard to understand, since the logic provides a top-down, or high-level, solution of the problem. By the interaction of the bit stream, the oscillator seeks for logical consistency, and once the consistency is reached, the oscillator stays in the state. The state can be a DC stable state, a periodic oscillation pattern, presence or absence of activities in the local loops, and the degree of stability of a state. I believe that the logic scheme is flexible enough to implement highly intelligent capability.

3.12 A State-Space Representation of Ringoscillator

A set of Boolean-levels of all the nodes of a ringoscillator, either of a single loop or of multiple loops, defines a *vector* in a binary-state space. As the state of the ringoscillator evolves, the state vector moves in the state space, and the trajectory is a *representation* of the ringoscillator operation. The trajectory is an abstract representation of an oscillation mode: the trajectory drawn by the moving-state vector threads a set of states in sequence, without specifying the initial state, which is a superfluous concept. A trajectory never has an intersection, since otherwise the state following after the intersection cannot be determined. Since there are a finite number of state points, 2^N in the N-node ringoscillator state space, and there is a trajectory that goes through any state point, a finite number of the representations covers the entire state space completely. A set of representations classifies the points of the state space into groups.

The examples of a three-member and four-member single-loop ringoscillators are shown in Fig.3.12.1. The three-member oscillator has a state space consisting of $2^3 = 8$ points. The eight points are arranged in two columns and four rows. The column index is the Boolean-level of node α, and the row indices are a composite of the Boolean-levels of nodes β and γ. The location in space is indicated by the three *coordinates*. The lower right-side corner, for instance, represents the state $\alpha = $ HIGH, $\beta = $ LOW, and $\gamma = $ LOW. A binary number is made by concatenating the column and the row indices. The column index is the MSB (most significant bit) of the binary number. The row indices are the two lower bits of the number. The binary number, or its shorthand decimal form, is used to identify the oscillator state. The states 1, 2, 3, 4, 5, and 6 appear in a single evolutionary development of the oscillator. The order of appearance of the states is $A_1, A_2, \ldots, A_6$. In this notation, the common principal character A indicates that all the suffixed A's belong to the same trajectory of an evolutionary sequence or to a single mode of oscillation. The suffix indicates the order of appearance of the state. If the states are identified by a decimal number as mentioned before, the order of appearance is 5-1-3-2-6-4. The trajectory is *discontinuous* in the space (the state transitions are not necessarily to the nearest neighbor points), but that does not matter. The states 0 and 7 are never reached by the evolutionary process. The two points X are $\alpha = \beta = \gamma = $ HIGH or $\alpha = \beta = \gamma = $ LOW. The two points represent the synchronous mode oscillation. The three-member ringoscillator has only two independent modes

Macrodynamics of Gate Field

of operation, since all the state space is covered by the two.

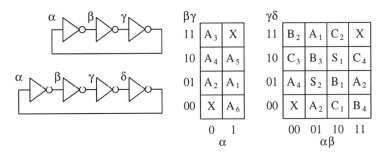

Figure 3.12.1 A state-space representation
of a single-loop ringoscillator

The state space of the four-member oscillator has $2^4 = 16$ points. The two diagonal points X represent the synchronous oscillation. As in the state space of the three-member oscillator, the synchronous operation points are at the ends of the diagonal of the space. The state space includes two steady-state points S_1 and S_2. In the single-loop ringoscillator, the steady-state points are all isolated points that cannot be reached from the other states. The balance of the points, twelve in all, are divided into three trajectories A, B, and C. In the oscillator operation in trajectory A, states A_1, A_2, A_3, and A_4 appear in succession, and after that state A_1 reappears and the cycle repeats. The oscillator works in the same way by trajectories B and C, but the Boolean-level profile circulating around the loop is different. A, B, or C is the *mode* of operation of the oscillator. Each trajectory is four states long, and each represents a steady-state oscillation.

Modes A, B, and C have the following node Boolean-level profile sequences:
A: (0111), (0100), (1101), (0001)
B: (1001), (0011), (0110), (1100)
C: (1000), (1011), (0010), (1110)
Modes A and C are essentially the same: the one mode is converted to the other by either complementing the Boolean-levels or by reassigning the node index. Mode B is different from mode A or C. There are three distinct dynamic modes of operation of a four-member ringoscillator, X, (A,C), and B, and a static mode S. Modes are originally the concept of the linear systems. If a ringoscillator consisting of a loop of four inverters is biased in the linear gain region, there are four modes of variation of the node voltage profile, and they are classified as follows (Shoji, 1992). The voltages of the four nodes 1, 2, 3, and 4 are, relative to the (unstable) equilibrium bias point voltages,

(I) All positive (or negative),
(II) Alternately positive and negative,
(III) The first node positive (or negative), the third node negative (or positive), and the rest zero,
(IV) The second node positive (or negative), the fourth node negative (or positive), and the rest zero.

Mode II grows into mode S, and mode I grows into mode X. Mode X has no oscillation of the spatial profile, and mode S has two oscillations in the spatial profile. Their correspondence to modes I and II is clear. Modes (A,C) and B have one oscillation in the profiles, and they correspond to modes III and IV in the linear region. The complexity in the correspondence is due to the phase of the single oscillation in the profile. Modes I, III, and IV are the decaying modes in the linear domain (Shoji, 1992), but in the idealized ringoscillator model we are studying now, all the modes grow and display their characters. The mode concept has been extended, although loosely, to the nonlinear circuit. Mode concept can be used to study any multiple-loop ringoscillators. As is observed from this example, a state space of a ringoscillator is useful in understanding the operation.

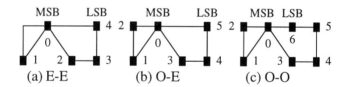

Figure 3.12.2 The structure of two-loop ringoscillators

Figures 3.12.2(a) to (c) show three examples of two-loop ringoscillators, having E-E, O-E, and O-O structures. They are the simplest nontrivial examples for each structure. Coordinates of a point in the state space are defined by assigning an order to the nodes. The node order for each oscillator structure is shown in Fig.3.12.2(a) to (c). In the state space of the E-E oscillator shown in Fig.3.12.3(a) the first two node states provide the four column indices, and the remaining three node states provide the eight row indices. There are total of thirty-two state points in the state space. In the state space of the O-E oscillator structure of Fig.3.12.3(b) the first three node states provide the eight column indices, and the rest the eight row indices. There are total of sixty-four state points. The states are identified by the decimal shorthand of the binary number made by concatenating the row indices to the column indices. States 0 to 9 are indicated in Fig.3.12.3(a). In Fig.3.12.3(a) and (b) the × signs indicate the states that eventually end up with the synchronous oscillation. The E-E (2-4) oscillator has two steady states, state

Macrodynamics of Gate Field 251

13 and 18, as shown by the dark squares in Fig.3.12.3(a). Although the stable states can be found by seeking for consistency of the Boolean-levels, the two stable states have significantly different dynamical properties, as we see later.

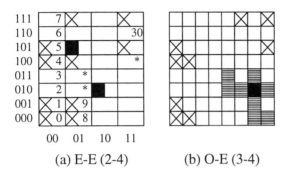

Figure 3.12.3 The state space of two-loop (E-E and O-E) ringoscillators

Many states that end up with the synchronous oscillation are clustered near the lower-left and the upper-right corners, but there are other states like 4, 5, 12, and 15 that are some distance away from the corners. Many, but not all, of them are the first neighbors of state (00000) or (11111). Even if the nodes are numbered approximately in sequence, there is no significant structural regularity in the state space. Clustering of similar state points are due to the generally small number of differences in the bit patterns, which are wiped out in the early evolutionary sequence of the oscillator. The unmarked states go through various evolutionary process, and their destination cannot be expressed adequately by the state space map. As we see later, all unmarked states in Fig.3.12.3(a) end up with stable state 18. In Fig.3.12.3(b) the × signs indicate the states that end up with the synchronous oscillation. Again, many of them are the first neighbors of the state (000000) or (111111). The dark square shows the single steady state of the O-E (3-4) oscillator, and the shadowed states end up with the steady state after several evolutionary steps. The unmarked states go through various dynamic states. Obviously, the state space map is not adequate to show the development of oscillator state adequately.

Figure 3.12.4 shows the more complete description of the oscillator operation the state transition sequence of the E-E (2-4) oscillator. Evolution of all thirty-two states of the oscillator is indicated in the figure. The thirty-two states are classified into three groups, A, B, and C. States 1, 4, 5, 8, 9, 12, 15 and 29 end up with the synchronous oscillation (group C), but how many steps are required to reach the destination depends on the state. State 13 is an isolated steady state (group B) that cannot be entered from any other state. The rest of the states go to the second-steady state 18 (group A). As I mentioned before, the two steady states are indeed quite different. Yet both states have their roots in the two mutually equivalent stable states of a binary latch, as the circuit structure became complex.

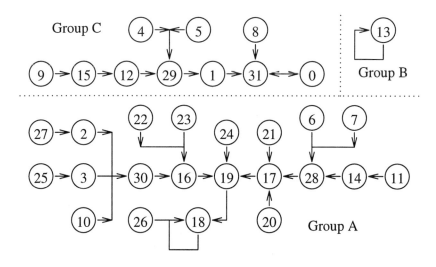

Figure 3.12.4 The state sequence of a two-loop (E-E) ringoscillator

The E-E (2-4) ringoscillator has dominant synchronous mode, whose oscillating period is 2, and there is no other mode of oscillation. The next complex O-E (3-4) ringoscillator shown in Fig.3.12.2(b) has the state transition sequence shown in Fig.3.12.5. State 42 is the single steady state of the oscillator. States 48, 49, 56, 35, 50, 51, 58, and 34 settle at the steady state in one to two evolutionary steps (group A). States 1, 4, 5, 8, 12, 47, and 61 settle at the synchronous oscillation in one to three evolutionary steps (group B). The rest of the states end up with the period 4 oscillation (group C). The oscillation is a repetition of the states, $43 \rightarrow 10 \rightarrow 62 \rightarrow 32 \rightarrow 43$, as is observed from the closed loop at the lower central part of Fig.3.12.5. The oscillating state can be entered from six states 2, 3, 33, 40, 54, and 55 that are directly connected to one of the four cyclic states. It takes up to six evolutionary steps to get into the steady-state oscillation having period 4. This is the dominant oscillation of this structure: there is no other oscillation, and there is a unique sequence of the states to produce the oscillation.

In Figs.3.12.4 and 3.12.5 the states are classified into three groups. It is interesting to investigate whether there is a structure whose entire state points belong to a single group. Two general conclusions can be drawn from this assumption:

Macrodynamics of Gate Field

(1) Since there must be the synchronous oscillation mode in any structure, the single group includes the two states, in one all the nodes are HIGH, and in the other all are LOW. This means that all the states end up with the synchronous oscillation mode after finite steps of evolution.
(2) Since there should be no stable state, no loop in the oscillator has even-numbered members. This statement means the following. If there is an even-numbered member loop, the alternating Boolean-level pattern of the loop can be set up such that the input of the multiinput gate at the confluent point from the loop disables the gate. Then the other loop is unable to oscillate even if the loop has odd-numbered members. The state is a steady state. This is a contradiction.

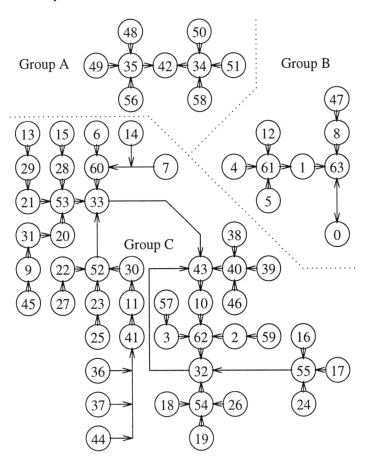

Figure 3.12.5 The state sequence of a two-loop (O-E) ringoscillator

The O-O (3-5) oscillator shown in Fig.3.12.2(c) has only one group in the state space. It is unnecessary to show the state-space map, since all the points belong to a single group. The state space has $2^7 = 128$ state points. The entire state transition sequence cannot be presented by a single drawing. Figure 3.12.6(a) shows the trunk structure of the transition sequence including the synchronous mode states, 0 and 127, and its connection to the three major branches, shown in Figs.3.12.6(b)-(d). The arrows in the branches are the *outputs* of it, and the branches are connected to the trunk at the arrow location, such that the state transition flow is from the branches to the trunk. This oscillator has only seven inverting components. Yet some states evolve nine steps before reaching the synchronous oscillation mode. Again, we observe that nearby states tend to cvolve together, as states 78, 79, and 80 of the trunk sequence, but the regularity is at the most tenuous, and the structure of the sequence is very complex.

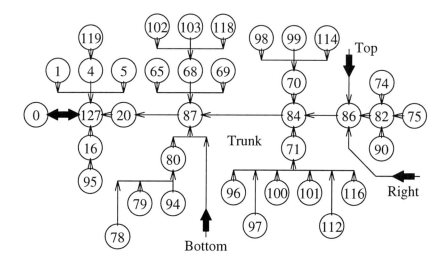

Figure 3.12.6(a) State sequence of 2-loop (O-O) ringoscillator

The examples show that the state transition sequence is the most complete method of describing the operation of a multiple-loop ringoscillator, but the problem with this representation is that the number of states increase exponentially with the number of the oscillator components. The combinatorial impossibility prevents general use of the transition-sequence diagram. It is quite difficult to draw the transition sequence of even modest three-loop oscillators. The heuristic method of the last section is then the most practical method for studying the properties of a multiple-loop ringoscillator.

The complexity of the state-transition sequence of even a simple multiple-loop ringoscillator is overwhelming. The complexity has not yet been well understood, and because of this lack of understanding the

Macrodynamics of Gate Field 255

ringoscillator structure has never been used practically.

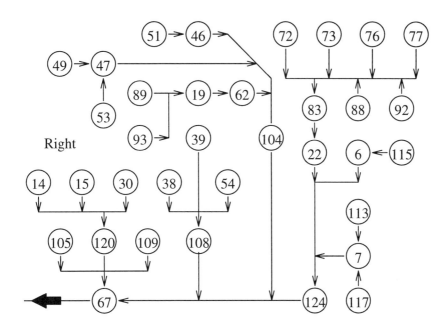

Figure 3.12.6(b) State sequence of 2-loop (O-O) ringoscillator

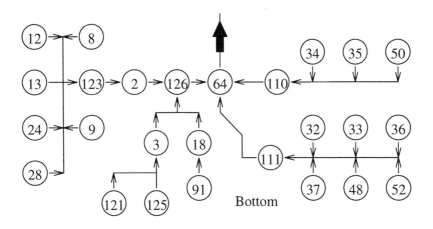

Figure 3.12.6(c) State sequence of 2-loop (O-O) ringoscillator

Yet the ringoscillator structure has great flexibility for displaying and processing data patterns. As the digital technology matures, practically all the conceivable circuit scheme has been exploited for application. In the future it is naturally expected that the capability of a ringoscillator may be applied to processing certain data that are not conforming to the standard digital electronics format. In the meantime, a ringoscillator is perhaps the closest model of the neuronic circuits of the central nerve system of animals. It is perhaps convincing that the great capability of the human brain originates fundamentally from the capability of the connected neuronic circuit making a loop. From this viewpoint, the study of multiple-loop ringoscillators may contribute to an understanding of the still mysterious function of the human brain.

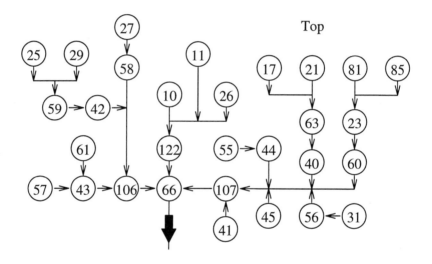

Figure 3.12.6(d) The state sequence of a two loop (O-O) ringoscillator

3.13 The Practical Significance of Ringoscillator Logic

The ringoscillator logic discussed in this chapter has the general structure shown in Fig.3.13.1. There are loops of gates, A, B, C, D, ..., and each loop has inputs shown by the small open circles and outputs shown by the small closed circles. Transfer of data from one location to the next of the propagation path can be effected either by the clock signal or by precisely delay-timed cascaded inverters. The loop nodes are set at the initial profile, and then the oscillator is released. The loops are coupled together by sharing the signal propagation path between the confluent point shown by the large open circle and the separation point shown by the large closed circle.

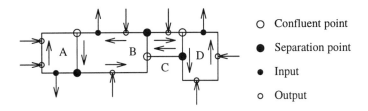

Figure 3.13.1 The structure of ringoscillator logic

Figure 3.13.2 shows the circuit of a node of ringoscillator logic. The data-storage medium at the node is either a node capacitor in the unclocked ringoscillator logic or a cascaded set-reset latch and a conventional latch in the clocked ringoscillator logic. In the former the node is forced by the external low-impedance voltage supply to either the Boolean HIGH or the LOW levels. When the switch connecting the node to the external voltage source is open, all at once the ringoscillator logic goes into the operation mode. The set or reset of the master latch of the clocked logic is carried out during the initial phase, when CK is low, and the input to the master latch is cut off. After the initial Boolean-level profile is established, the oscillator is either unclamped or clocked, and the data profile circulates around the loop.

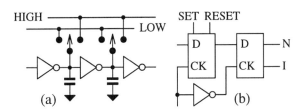

Figure 3.13.2 The structure of nodes

Ringoscillator logic is different from conventional logic, which can be decomposed into the elementary NAND or NOR gates. The logic is suitable for executing rather complex operations in a simple structure, and as such, a lot of design ingenuity is required. For some applications the logic is quite efficient. As an example, let us consider the following. Suppose two integer numbers, N_1 and N_2, such that $N_1 < N_2$, are given. It is often necessary to determine whether the numbers have a common integer factor. This operation can be carried out by ringoscillator logic as follows. We connect an N_1 member loop and an N_2 member loop by an OR gate. The unit of the loop is a noninverting master-slave clocked latch. The initial condition is set up as follows: all the nodes, except for the node of the smaller loop immediately upstream of the OR gate, are set at the Boolean LOW level. The special node of the small loop at the gate's upstream is set at a Boolean HIGH level. If the ringoscillator logic is clocked, the following things happen. If N_1 and N_2 have no common factor except 1, all the nodes of the oscillator are filled up by the Boolean HIGH level, as shown in Fig.3.13.3(b). The oscillation stops. If they have a common factor other than 1, the oscillation continues forever. This is a very simple structure to carry out the complex mathematical operation to determine whether N_1 and N_2 have a common factor.

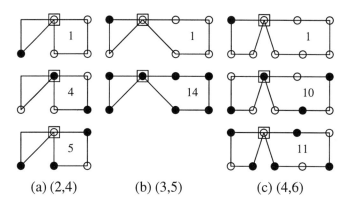

(a) (2,4)   (b) (3,5)   (c) (4,6)

Figure 3.13.3 Detection of a common factor

The dominant trend of present digital electronics is to execute all the complex-logic operations by clocked sequential-logic circuits. A clocked sequential-logic circuit has a bank of memory devices that holds the digital input data to be processed by a circuit built from combinational-logic gates. The input data are delivered to the combinational logic circuit at a clocktick. The logic answers from the combinationa-logic circuit are captured by the latches of the second bank of memory devices at the second clocktick. The digital data in the second bank are then transferred back to the first bank, and the operation repeats. The control of the memory banks is carried out as rapidly as the device technology permits, by using the edge-triggered data transfer scheme (Mano, 1991). The memory device is some variation of latch that has the capability of Boolean-level determination and retiming for recovering a full level-to-level, crisply switching digital-signal waveform from the generally corrupted output waveform from the combinational-logic circuit. The signal-quality improvement is a feature that exists only in a digital circuit and not in analog circuit.

A synchronous sequential logic circuit requires a clock. The clock is generated outside, to match the capability of the logic circuit, and it drives the latches to deliver and capture the signal. Except for the combinational-logic-gate circuit, the clock and the momory are the two essential ingredients of a synchronous sequential logic circuit. This general characterization provides an assessment of the features and the practical applications of the ringoscillator logic circuit proposed in this chapter. A ringoscillator obviously has the clock, which is self-generated, and therefore it matches the circuit capability. As for the memory, the discussion of Section 3.05 shows that a ringoscillator does indeed have a built-in memory: a section of cascaded inverting gates provides delay, the essential component of memory (Mano, 1991; Nagle, Carrol and Irwin, 1975). This type memory is an attribute of an asynchronous sequential-logic circuit. Then the ringoscillator logic is a member of the sequential-logic circuit family, but it stands somewhere between the synchronous and asynchronous versions. Because of the intermediate character, it has number of unique and peculiar features.

Let us investigate the ringoscillator logic by focusing attention on how peculiar they are. Three features are the most outstanding:

(1) The *memory* of the ringoscillator logic is the region on the ring where the Boolean-levels of the nodes make an alternative HIGH and LOW pattern, H-L-H-L-H-L.... The alternative pattern region is terminated by H-H or L-L pattern at the ends, and the H-H or L-L pattern moves with time. This is a difference from the bank of latches of a conventional sequential logic circuit, which stays at the same location of the circuit. At the confluent point the width of the section changes. The memory-storage location moves with time, and the length changes. The memory is a type of dynamic memory that refreshes as it operates. In a long time the width of the region may increase or decrease by the natural decay process (Sections 3.06 and 07).

(2) The clock frequency is not constant. Depending on the data supplied to the ringoscillator logic circuit, the logic circuit itself determines the clock frequency. The clock may even stop if a steady state is reached. The ringoscillator logic belongs structurally in the general category between a synchronous and an asynchronous sequential logic circuit, but it positively uses the undesirable feature of asynchronous sequential logic, the instability, to execute the logic. If used properly, this would be a unique, efficient, and convenient feature. Features (1) and (2) suggest that the logic circuit is better integrated as a whole than a conventional logic circuit, in the sense that all the gates participate in the operation, and that means the circuit can be quite hardware efficient for certain purposes.

(3) The operation of the ringoscillator logic circuit has an *objective*, and the circuit operation may be considered to be as *effort* for accomplishing the objective. This is most clearly seen from the examples of Sections 3.07 and 3.13. The problem is given all at once as the

initial data pattern on a ringoscillator, and as the oscillation proceeds, certain features are dropped and the final logic level (in case N is even) or the final oscillation state (in case N is odd) is reached. This is a highly intelligent capability. Since many problems that a computer must solve are consistency seeking, this is a quite useful feature. One obvious application of ringoscillator logic is radix-independent logic operation. Ringoscillator logic has the gear-operation mode (Section 3.09). The number of teeth in a gear is the radix N. The N numbers can be stored in a loop directly. Manipulating the number can be carried out by interaction of the loops as in the example of Fig.3.13.3. The gear-based computing machine of the last generation which introduced modern information processing for the first time, such as the telephone switching machine and the mechanical calculator can be upgraded by using the ringoscillator logic. This type of operation is believed to be useful in executing a complex function at the highest circuit speed the device technology allows. The isolated pulse removal is one such example at the high-speed limit.

3.14 An Asynchronous Multiloop Ringoscillator

Understanding the operation of a synchronous multiloop ringoscillator in the idealized and simplified gate-level model is still quite complex, but it is manageable. The complexity can be handled by using computer experiments. If transfer of excitation from one location to the next takes a range of delay times not regulated by the clock period or by precision of the circuit parameters, the ringoscillator operation becomes quite complex. Yet such a ringoscillator logic can be useful (Section 3.07). The most difficult is that the universal discrete time points, when the Boolean-levels of all the nodes are definite, cease to exist. How do we understand the circuit in this case? We make a list of the complications. There are three significant effects that make the problem hard, whose clarification provides a lot of insight:
(1) The phenomena within a single propagation path,
(2) The logic operation at the confluent point of more than one paths, and the phenomena at the branch point of the path, and
(3) How to represent the state of a ringoscillator, since no universal sequence of time points when all the oscillator nodes have definite Boolean-levels.
Problem (3) is practically the most troublesome, since it defeats the conventional gate-level approach. We rely on the circuit-level analysis, with the gate-level analysis as a guide, and we introduce some simplifications. Let us consider the two waveform-related issues of the signal path and the confluent point (1) and (2) as the gate-level problem and consider how to correlate the events of the paths and of the gates.

The inverter chain that connects the output of the first combinational gate at a confluent point to the input of the second combinational gate (the second gate may be the first gate itself) has delay time T_D. The delay time is defined by setting a logic threshold voltage: the following analysis is classical, or the node waveforms are assumed to have zero rise-fall times. The signal at the output of the first gate appears at the input of the second gate time T_D later, in the same polarity or in the inverted polarity, depending on the number of inverting gates in the path. The Boolean-level of the input of the second gate at time $t+T_D$ equals the polarity-adjusted Boolean-level at the output of the first gate at time t. In this simplest model, the gate chain executes simple time axis shift transformation and possibly inversion. Let the first gate output make a transition at $t_0, t_1, t_2, \ldots$. Let the even-numbered suffix time point be an upgoing transition, and the odd numbed suffix time point be a down-going transition. The transition edges arrive at the input of the second gate at

$$t'_0 = t_0 + T_D \quad t'_1 = t_1 + T_D$$
$$t'_2 = t_2 + T_D \quad t'_3 = t_3 + T_D$$

and so on, with the adjusted transition polarity. They are the first approximations.

The inverter chain has the mechanisms to make the input upgoing delay and the input downgoing delay different. The difference is given by adding correction $T(L \to H)$ or $T(H \to L)$ to each delay. By including the correction the transition edges arrive at the input of the second gate at

$$t'_0 = t_0 + T_D + T(L \to H) \quad t'_1 = t_1 + T_D + T(H \to L)$$
$$t'_2 = t_2 + T_D + T(L \to H) \quad t'_3 = t_3 + T_D + T(H \to L)$$

This is the second approximation.

The trailing edge of the last excitation and the leading edge of the present excitation interact, and the interaction reduces the width of the pulse in between (Sections 2.17, 2.18, and 3.07). The effect is cumulative over all the gates in the chain. The transition edge at t_0 tends to pull the transition edge at t_1 and therefore a correction, which is a function of $t_1 - t_0$, must be added to the estimate. Then the transition edges arrive at the input of the second gate at

$$t'_0 = t_0 + T_D + T(L \to H)$$
$$t'_1 = t_1 + T_D + T(H \to L) - \tau_1(t_1 - t_0)$$
$$t'_2 = t_2 + T_D + T(L \to H) - \tau_2(t_2 - t_1)$$
$$t'_3 = t_3 + T_D + T(H \to L) - \tau_3(t_3 - t_2)$$

where $\tau_i(**)$ is a function of the argument, generally an exponential function (Section 2.18). The set of equations define the time domain transform from $(t_0, t_1, \ldots, t_N)$ to $(t'_0, t'_1, \ldots, t'_N)$. The τ function depends on $t_i - t_{i-1}$. This is an approximation that is valid only if all

t'_is can be determined from t_i. If none of the intervals between the transition edges $t'_m - t'_{m-1}$ is small, the transformation gives the arrival time of the transition edges at the input of the second gate correctly. If any of the intervals between two transition edges becomes zero or negative, the narrow feature between the edges disappeared. Suppose that $t'_m - t'_{m-1} \leq 0$ (pulse 1). Then the pulse between the edges at t'_{m-3} and t'_{m-2} (pulse 0) and the pulse between t'_{m+1} and t'_{m+2} (pulse 2) become consecutive. Then the interval between pulse 0 and pulse 2, $t'_{m+1} - t'_{m-2}$, is not an accurate measure to determine the location of the edge at t'_{m+1} in the subsequent development, since there can be the half-decayed waveform of pulse 1 in the interval between t'_{m-2} and t'_{m+1}. The method of computation of the edge location must be readjusted if any of the isolated pulse decays and disappears. By this mechanism the propagation path may decrease the number of the isolated pulses, as schematically shown in Fig.3.14.1.

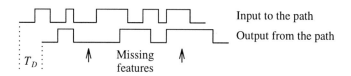

Figure 3.14.1 Node waveforms at the confluent point

A spurious pulse is an essential byproduct of logic operation, but then how does its decay affect the other pulse timing? In a conventional synchronized sequential-logic circuit the number of the spurious pulses is limited, but in a ringoscillator logic circuit there can be any number of narrow pulses, which are set up as the initial Boolean profile. A ringoscillator logic is an useful vehicle to study the effect. Figures 3.14.2(a)-(d) shows various modes of decay of an isolated pulse propagating through a symmetrical inverter chain. The FETs are modeled by the gradual-channel, low-field model, and the parameter values are

$$B_N = B_P = 1 \quad V_{DD} = 1 \quad \text{and} \quad V_{THN} = V_{THP} = 0.1$$

The node voltage waveform at the input is shown by the solid curve. The input voltage waveform consists of two wide pulses whose width is 12 (in normalized time unit), and they are separated by 18. A narrow pulse between the two wide pulses, and for Figs.3.14.2(a)-(c) the width is 6, and for Fig.3.14.2(d) it is 5. The narrow pulse is centered between the two wide pulses in Fig.3.14.2(a), located a little left, or before the center in Fig.3.14.2(b) and a little right (after) in Fig.3.14.2(c). The node-voltage waveform at the output of the sixth inverter from the input was plotted.

The numerical analysis shows the following. If the narrow pulse is centered, its amplitude decreases because the LOW logic level moves up and the width narrows, but otherwise the pulse remains centered until it collapses and merges the two wide pulses, as observed from Fig.3.14.2(a). If the pulse is shifted to either side, the narrow downgoing pulse between the pulses collapses first, and the narrow pulse merges the wider pulse, as it is observed from Figs.3.14.2(b) and (c). After a long time all the three pulses merge together. If the narrow pulse is quite narrow, it decays by itself, by the interaction between its own leading and trailing edges, as observed fron Fig.3.14.2(d).

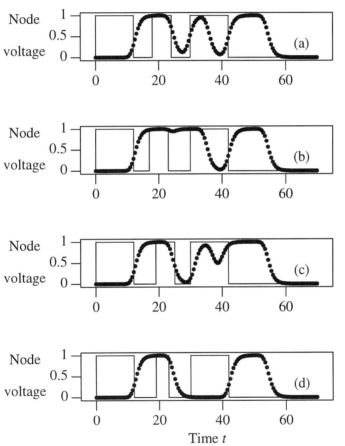

Figure 3.14.2 Various modes of decay of an isolated pulse

The waveforms at the confluent point are a time-axis transformed (locally extended or shrunk) version of the original waveform, including the possibility of some pulse elimination, as schematically shown in Fig.3.14.1. The feature elimination does not happen, if the signal input to the path does not have features that are too narrow (typically narrower than five times the gate-delay time). When a coupled ringoscillator arrives at an apparent steady state, this is what is observed. Here a *steady state* never means the absolutely stable state as in the clocked circuit, but it is a state in which the features have exponentially long life, over many cycles of oscillation.

As for problem (2), the multiinput gate at the confluence point can be understood from the conventional digital circuit theory. The gate may produce various responses. The possible cases are classified as follows:
(a) If the gate is disabled, the output is in the default level. For instance, if it is a NAND gate, the output is the default HIGH level. If one input disables the gate and if a transition edge arrives at the other input, the edge never passes the gate. By this mechanism the gate decreases the number of the pulse features.
(b) If the gate is enabled over the period of time by one of its inputs, the gate works as an inverter. In this case the number of the waveform features of the active loop remains the same.
(c) At the branch point of a path, the number of transition edges increases.
(d) If the two input signals having opposite transition polarity arrive at different times but consecutively, the gate may produce an isolated pulse. The number of transition edges remains the same.

The first three mechanisms (a) to (c) are the essential operational modes of a logic gate, and no further discussion is necessary. Whether or not the fourth mode is the designer's intention depends on the case. A narrow pulse is generated if the operation includes a do-undo process pair that cannot be avoided if stepfunctions having different transition polarities are combined at the confluent point. Suppose that there is a two-input NAND gate in a logic circuit, the input A is in the LOW logic level, and the input B is in the HIGH logic level. Both inputs are driven by logic gate chains. If input B gets a HIGH to LOW stepfunction first, and then input A gets a LOW to HIGH stepfunction, the output of the NAND gate remains at the disabled default HIGH logic level all the time. If the order of the signal arrival is reversed, the output is in LOW logic level during the interval between the time of arrival of the upgoing stepfunction to input A to the time of arrival of the downgoing stepfunction to input B. In either case the output of the NAND gate settles at a HIGH logic level ultimately, but in the first case it occurs without generating an extra isolated pulse, and in the second an isolated downgoing pulse is generated.

Issue (3) of an asynchronous ringoscillator is how to model it as a manageable circuit. One way to extend the gate-level approach to the asynchronous ringoscillator is to proceed as follows. The output node of a combinational gate at the confluent point makes a transition at $t = 0$. At this time we consider that the signal goes into a black box that represents the signal path. After the delay time of the path, the time-axis transformed signal appears at the input node of the second combinational gate, and it is processed by the second gate. We adopt first a very restricted model, in which only the input and the output nodes of the combinational gates at the confluent points are observable. Since this simplifies the model to quite a significant extent, the *skeleton* model by the gates only of the multiple ringoscillator can be studied for all the possible timing relations at the combinational gate inputs. This study provides the over all framework of the development of the system. Once the framework is established, we substitute the black box by the signal path, and determine the evolution of all the nodes. The signal that goes through the gate chain signal path can be determined by the method described before. This method is, in essence, to separate the ringoscillator into the delay elements and the logic circuits.

Digital information is represented by the two types of stepfunctions. If a stepfunction disappears, the information associated with it is lost. The studies of Chapter 3 showed clearly that decay of a stepfunction occurs by two fundamentally different mechanisms. The first decay mechanism is the valid operation of a logic gate that is in the disabled state. The stepfunction that propagates down the logic chain stops at the disabled gate completely. In this mechanism, the stepfunction decays at a specific spatial location, and the number of the stepfunction edges decrease by one at a time. The second mechanism is the decay of a stepfunction pair having the opposite transition polarity, or reduction of the number of stepfunction edges by two. This mechanism involves details of the electrical characteristic of the gate chain. The pulse-decay process spreads out over many gate or inverter stages, and the exact location cannot be specified.

3.15 The Precision of an FET Model and Simulator

We saw that the circuit simulator to design an ultrahigh-speed circuit and the asynchronous ringoscillator logic circuit must be able to determine the node-voltage waveforms accurately. From the accurate node waveforms the delay times are determined and the glitch pulse is evaluated. The accuracy of a simulator depends on how the following four issues are handled:
(1) Accuracy of the equivalent circuit representing the circuit on the silicon chip,
(2) Accuracy of the numerical analysis of the simulator,
(3) Accuracy of the model of the active and passive components, and
(4) The node waveforms affecting the effective device characteristics.

(1) Accuracy of the equivalent circuit depends on the way the real circuit on a silicon chip is coded to the simulater code as a discrete component circuit. Resistive and capacitive components can be included without fundamental difficulty. Inclusion of inductance creates many fundamental problems, as I discussed in my last book (Shoji, 1996). The

state-of-the-art high-speed circuit is still simulated by using a resistive-capacitive equivalent circuit and only to a limited extent by using a lossy transmission-line model including inductance. This simplification is acceptable, so far as the circuit's overall size scales down to the same ratio as the switching-delay time and the clock period. If 1GHz clock circuit is limited to a linear dimension of about a 100-micron-sidfe square, the signal propagates from a corner to the other corner in 0.06 picoseconds, and this time is small enough to be negligible. This is not the case of a ULSI, like a microprocessor. Here we consider that inductance is not included. There are enough problems already in the RC-only circuit model.

Modern CMOS technology allows more than one level of metal interconnection. The capacitance between the overlapping metal levels must be included in the equivalent circuit, if the node is sensitive. Then the equivalent circuit cannot be drawn until the layout of the circuit is finished. A *conceptual* design exercise before the final design becomes difficult. The capacitance to the less conductive levels, such as to gate polysilicon (typically a few 10s of ohms per square) and to the substrate (typically several kiloohms per square) must be substituted by a capacitance-resistance series connection, but that requires layout information as well. Exactly how the circuit is laid out is uncertain at the time of the conceptual circuit design.

An essentially new problem appears if multilevel interconnect interaction is included. A nodes of the circuits on a chip that have capacitance to the other node of the same chip are usually active nodes, which have their own drivers. The nodes may change state during the critical-circuit operation. The capacitive coupling introduces essentially synchronous *noise* to the circuit, which may be mistaken by the circuit as a part of the essential signal. This effect is inevitable in a clocked, synchronous digital circuit, since any signal that exists in the circuit is a version of the clock modified by the enable-disable signals (except for those arriving from the outside asynchronously, even they are versions of the high-level system clock). Quite a lot of uncertainty exists in the high-speed equivalent circuit.

Yet the most important issue of the equivalent circuit is how to isolate the critical path properly from the host circuit for delay simulation. Simultaneous input switching of a gate in the same transition polarity affects the delay of the gate seriously, and sequential switching of the inputs in the different transition polarity may create an isolated pulse or hazard. Then extraction of critical path demands careful consideration not to miss the delay-critical simultaneous/near simultaneous switching events. If the entire circuit can be simulated at once, the problem is solved, but that is not always possible because of the complexity. A combinational CMOS logic gates like NAND and NOR gates generate an isolated pulse or glitch during the interval when all the inputs take the logic level that enables the gate. The gate goes into the state and gets out of the state by many different sequences of input switching. Glitch generation has huge varieties. This makes identification of the critical-path-equivalent circuit quite difficult. Setting up the equivalent circuit is hard because there are too many cases for a human mind to deal with. Since glitch generation depends on the input signal timing to a multiinput gate, there can be combinationally many cases that must be examined at the time of design verification. The combinational multiplicity of verification is unfortunately essential, and there is no way out. The present practice is to choose the equivalent circuit intuitively, and there is no systematic method of choice. Some improvement is required by a CAD that provides some guide to this difficult problem. One way to avoid this problem is to use the simplest multiinput gate such as NAND2 and to time the input signal properly. Due to the waveform distortion by glitch generation, the circuit timing becomes uncertain through the accuracy of the simulator. This effect is aggravated by its data dependence.

(2) The accuracy of the numerical analysis depends on several factors. In the type of circuits that use FETs deep in the saturation region most of the time for instance, operational amplifiers and analog circuits the DC bias point setting is often quite sensitively dependent on the parameter values. The initial condition of digital latches is uncertain, unless specified by the circuit structure itself or by the circuit simulator code. A modern circuit simulator is always designed to determine the initial DC bias-point setting by one algorithm or the other, intending not to abort simulation. The operating point may or may not be realistic, since there can be more than one stable bias point. A metastable bias point may be wrongly chosen as the initial state, or a totally wrong initial condition may be set in the uninitialized latches. Although a simulator that is always able to set the initial condition is convenient for general design purposes, the result can be quite misleading. A thorough check is required.

If a critical path delay is simulated by enabling all the gates in the path by the Boolean-level voltage, simultaneous switching of the gate is never reflected to the simulation result. If simultaneous switching is to be examined in an equivalent circuit including all the signal paths starting from a set of the clocked latches, a small error in the relative timing of the paths is amplified at the gate where multiple signals from the source latches are combined. If all the signals have the same transition polarity, the gate delay critically depends on the signal-arrival timing. If the signals have different polarities, an isolated pulse may or may not be generated, and if it does, its width affects the waveform distortion in the downstream. Once an unwanted isolated pulse is launched, the simulation accuracy in the downstream nodes becomes worse. The two mechanisms increase delay uncertainty to quite a significant degree.

The requirement for accuracy in high-speed circuit simulation is quite stringent, since it is often necessary to determine not a delay time but a difference in the delay times of two signal paths. This is necessary to control the race condition and the waveform distortion. The difference of two numbers each having errors has a higher percentage of errors than the original two numbers. This problem can be partially compensated by constructing the parts of the two circuits identically, including the layout structure. Then the difference is pushed to the dissimilar part. Since the identical part must be identical in every way, it is not always easy to layout, but this method certainly decreases the heavy dependence on simulation accuracy.

(3) The accuracy of the model and the parameters of the active and passive components affects the simulation accuracy through the following mechanism. FET is a nonlinear device. In a large-signal delay-time analysis of a logic circuit, the nonlinearity makes some device parameters more important than the others. Figure 3.15.1(a) shows a typical NFET drain current I_D-drain voltage V_D characteristics.

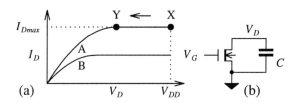

Figure 3.15.1 An FET model affecting simulation accuracy

We consider the NFET discharges capacitor C by the circuit shown in Fig.3.15.1(b). The capacitor is initially charged to V_{DD}. In Fig.3.15.1(a), A is the I_D-V_D characteristic when the gate voltage V_G equals V_{DD}, and B is for V_G lower than that. If V_G makes an instantaneous transition from 0 to V_{DD}, the NFET assumes the I-V characteristic A immediately after the transition. The capacitance C of Fig.3.15.1(b) discharges at a constant rate I_{Dmax}/C (volt/sec) following the path shown by the arrow, from the initial bias point X to Y. This process can be determined practically by I_{Dmax} only. To determine a single FET parameter accurately is an easy task. The delay time of a gate is determined more by the early phase-switching waveform than by the later-phase waveform. The FET characteristic in the Y to O transition phase does not matter significantly to the part of the waveform affecting the delay-time accuracy. In this switching mode the FET model-accuracy requirement is not hard to satisfy. In the other case, if the signal amplitude is small, as in a linear circuit, the accuracy is again dependent on only a few parameters, like the gain and the time constant.

If V_G makes a gradual transition, the output node discharge waveform is determined not by the single I-V characteristic A but by many I-V characteristics such as B, for the lower-gate voltages. Then the requirement for the FET model becomes qualitatively different: many more parameters other than I_{Dmax} must be specified accurately. This crucial difference has its origin in the two modes discussed in Section 1.05, the static and the dynamic switching modes of an inverter: the ideal dynamic switching mode depends only on a few inverter parameters, but the ideal static switching mode reflects the input waveform to the full extent, through which many FET parameters get involved. As the circuit speed becomes higher, the gates are more lightly loaded, and the gate's switching mode becomes more static switching. The reason that this effect becomes significant at the ultrahigh clock frequencies is as follows. In a digital circuit, the clock signal usually has the shortest rise-fall times. The other signals may be considered as the enabled or disabled versions of the clock. In a low clock-frequency circuit, the signals that enable the gates arrive well before the signal, which may be considered as a modified clockedge. Then the fastest signal drives a gate, which has some time to switch the output node since the output node of the gate is likely to be loaded significantly, to carry out complex signal processing. Then the logic gate is more likely to be in the dynamic switching mode. The signals produced by the gate are further processed in the logic downstream, but the signal that further modifies the signal from the upstream is again generated much earlier. This is the issue of logic-design style and usage of the gate types. If the device technology does not improve proportionally to the required speed, the circuit-design style changes, and the gate usage is restricted. In a high-speed circuit there is no flexibility. The clock and its enable and disable signals are generated at about the same time. The clocks do not have a significantly shorter rise-fall time than any other signal. The input and the output switching waveforms overlap more in high-speed circuits. Then the gate operation becomes more dependent on the input waveform including the clock signal, a definite signature of the static switching mode. From this viewpoint a static switching is a partial parallelization by overlapping the signal processing in a chain of logic gates, and that is the reason that the static switching mode appears at the highest circuit speed.

To conclude this section, I summarize as follows. In the high-speed circuits two issues are fundamentally important:
(1) Interaction of signals originating from a set of latches that are clocked by the common clock, and
(2) Generation and annihilation of an isolated pulse, and the associated distortion of the node waveforms.
CAD tools that specifically address the two problems are necessary.

As for issue (1), the CAD tool is not to carry out a circuit-level simulation. Rather, it identifies the signal paths that must be included in the critical path simulation. The source and the destination latches must be included. It is an aid to drawing a proper equivalent circuit diagram. The circuit diagram depends on many run-time parameters, most likely, on the data pattern, and the proper equivalent diagram is required for each case. The CAD tool scans the circuit by using a simple delay computation algorithm, such as those used in the gate-level simulator, and determining whether a signal that drives a gate arrives much earlier than the other signals. If the signal does, the signal may be considered to be an enable-disable signal, and the circuit upstream of the gate can be eliminated by setting the node level either Boolean-level HIGH or LOW to enable the gate. Since the spurious part of the circuit is eliminated from the simulation, it takes less time to run a simulation. Yet the equivalent circuit includes all the details of simultaneous switching of the gates. Such a CAD tool is useful not only for equivalent circuit extraction but for the circuit design and conception. Circuit-level simulation can be carried out by a conventional circuit level simulator that is already available, after proper equivalent circuit is drawn.

As for issue (2), the CAD tool identifies the location of the circuit where an isolated pulse generation cannot be eliminated. To study the node waveform distortion, the tool must be a circuit-level simulator to determine the node waveform accurately. Generation of an isolated pulse by a gate, especially of a narrow spurious pulse, has not been well understood. A detailed study of the glitch-generation mechanism by the circuit theory is a prerequisite to developing such a CAD tool.

3.16 Conclusion

An electronic circuit processes information represented by electrical variables. The complex information that a digital circuit processes is composed of elementary components. Digital-signal elementary components are a pair of digital excitations that are stepfunctions one upgoing and the other downgoing. They are localized spatially in the digital circuit.

From the numerous examples presented in this book, the following conclusions can be drawn. The fundamental reason that a digital circuit and a digital excitation have the aspects that conform to the laws of mechanics is that the elements of the digital signal are an undividable whole that is localized and that carries its own identity. As long as the identity is sufficient, the excitation can be treated as a classical particle. Digital excitation is quite elementary, however, to the degree that it cannot be sufficiently labeled. Here come the quantum-mechanical aspects. Digital electronics and quantum mechanics deal with objects that are too elementary to carry their own complete identification. In quantum mechanics, the object is reached on repeated division of a macroscopic substance, and the division is a process of miniaturization and simplification. In digital circuits, the simplicity and elementarity of the excitations are reached by dropping the physical meanings of many variables that are still measurable by ordinary physical test equipment or by analog measurements. Dropping the attributes can be carried out consistently by restricting the test procedure allowed in the gedanken experiment, as shown in Chapter 2. This procedure has clear physical meanings. In this sense a digital circuit is an embodiment of the quantum world, as the hidden variable interpretation of quantum mechanics dictates (Bohm, 1951). In spite of the difference, both processes ended up with objects that are essentially quite similar. Between the two similar objects, it is no surprise that a number of functional parallels can be found.

A conventional digital circuit operates in a heavily compromised mode, neither a classical nor a quantum-mechanical mode. A better digital-circuit engineering hides its quantum-mechanical nature more. Yet the fundamental quantum-mechanical nature of a digital circuit shows up in the compromised engineering product, and that creates a lot of confusion and problems. The present work attempts to clarify the source of this confusion. Once the connection between the two distant branches of natural science is established and the fundamental nature of digital circuit is well understood, we have no more confusion, and we have a rich source of ideas to design high-speed digital circuits. This is the most fascinating prospect that I, a practicing IC designer, face. This book was written to share my excitement with the readers.

3.17 The Future Direction of Digital-Circuit Research

In the four books I have written in the field of digital-circuit theory, I showed the following:
(1) A digital circuit is an interesting and rich subject for physics research (Shoji, 1987).
(2) Digital-circuit operation is described by a sequence of transitions among the thermodynamic states of the circuit (Shoji, 1992).
(3) At the ultimately high-switching speed where the interconnect inductance becomes significant, the circuit operation is described by an exotic theory that allows time to flow forward and backward. In spite of the unconventional feature, the theory is still a natural extension of the low-speed theory of the thermodynamic state transitions (Shoji, 1996).
(4) In this work a digital signal is considered as an excitation in the circuit that behaves like classical and quantum-mechanical particles moving in the circuit.

These four new viewpoints are rational but are not conventional, and they have not yet been universally accepted. Yet they are the basis for the modernization of electronic digital-circuit theory to the twenty-first century.

Standing on the ground, where shall we go? The invention of a new circuit is the invention of new component connectivity. Creating a new circuit connectivity for executing new and desirable functions simply and effectively has always been an art rather than a systematic science. A systematic theory that correlates the component connectivity to the circuit functionality must be built. This problem directly faces the combinatorial complexity. How to squarely combat this formidably large and difficult problem is the challenge that wait us in the future. A useful circuit is ever more needed in integrated-circuit technology. Just the number of components is finite and not very many, the number of useful circuit is also finite. We need to discover all the useful circuits. How do we do that? Do we generate all the possible circuits by connecting components in all possible ways and then filtering the useless circuits out? Or is there a better and more efficient way?

The connection between component connectivity and circuit functionality will obviously lead us to a still more difficult problem that is within the domain of electronic circuit theory, I believe: how to model human brain, which is essentially a gigantic digital circuit. To approach this difficult and confusing problem in the boundary region between natural science and philosophy, electronic-circuit theory has one undisputable advantage: the theoretical framework is so realistic and precise that unsubstantiated speculation and handwaving are never allowed. the steady accumulation of knowledge about the mode of operation of the complex circuit and perhaps even freaky circuit (Shoji, 1992) will one day open a road that reaches the model. From this viewpoint, it is my hope that electronic-circuit theory will become the meeting ground of physics and the information sciences, including psychology. Once that integration is made, natural science will reach one step higher level or be dialectically synthesized.

References

Aitchison, J.J.R. (1982), Gauge theories in particle physics, Bristol: Adam Hilger

Bohm, D. (1951), Quantum theory, Englewood Cliffs, NJ: Prentice-Hall

Brews, J.R. (1981), Physics of the MOS transistor, Applied solid-state science, supplement 2A, NY: Academic Press

Carey, R., and Issac, E.D. (1966), Magnetic domains and techniques for their observation, NY: Academic Press

Chaney, T.J., and Molnar, C.E. (1973), Anomalous behavior of synchronized and arbiter circuits, IEEE Transactions on Computers, C-22 (April) 421-422

Freedman, D.Z., and van Nieuwenhuizen, P. (1985), The hidden dimensions of space-time, Scientific American, 252 (March) 74-81

Glasser, L.A., and Dopperpuhl, D.W. (1985), The design and analysis of VLSI circuits, Reading, MA: Addison-Wesley

Goncalves, N.P., and de Man, H.J. (1983), NORA: A racefree dynamic CMOS technique for pipelined logic structures, IEEE Journal of Solid-State Circuits, SC-18 (June) 261-268

Heller, L.G., Griffin, W.R., Davis, J.W., and Thoma, N.G. (1984), Cascode voltage switch logic: A differential CMOS logic family, ISSCC 84 Digest, (February) 16-17

Heitler, W. (1945), Elementary wave mechanics, Oxford: Clarendon Press

Kang, S.M., and Leblebici, Y. (1996), CMOS digital integrated circuits: Analysis and design, NY: McGraw-Hill

Landau, L.D., and Lifshits, E.M. (1958), Quantum mechanics, London: Pergamon Press

Mano, M.M. (1991), Digital design, Upper Saddle River, NJ: Prentice-Hall

McCluskey, E.J. (1986), Logic design principles, Englewood Cliffs, NJ: Prentice-Hall

Nagle, H.T., Carrol, B.D., and Irwin, J.D. (1975), An introduction to computer logic, Englewood Cliffs, NJ: Prentice-Hall

Penrose, R. (1994), Shadow of the mind, Oxford: Oxford University Press

Prigogine, I. (1961), Introduction to thermodynamics of irreversible processes, NY: Interscience, John Wiley & Sons

Ridley, B.K. (1963), Specific negative resistance in solids, Proceedings of the Physical Society, 82 954-966

Shepherd, G.M. (1994), Neurobiology, NY: Oxford University Press

Shoji, M. (1987), CMOS digital circuit technology, Englewood Cliffs, NJ: Prentice-Hall

Shoji, M. (1992), Theory of CMOS digital circuits and circuit failures, Princeton, NJ: Princeton University Press

Shoji, M. (1996), High-speed digital circuits, Reading, MA: Addison-Wesley

Shoji, M., and Rolfe, R.M., Negative capacitance bus terminator for improving the switching speed of a microprocessor databus, IEEE Journal of Solid-State Circuits, SC-20 (August) 828-832

Stapp, H.P. (1993), Mind, matter and quantum mechanics, Berlin: Springer-Verlag

Index

autonomous region (of switching): 29-31, 34, 38
background (of circuit): 12-13, 17
backkick (of inverter switching): 69-70
backtracking (causality): 29-30, 34, 36, 38-39
barrier (in gate field): 5
bidirectional (signal propagation): 9-10, 79
Boson (circuit model): 178, 183, Section 2.20
collapsable current generator model (of FET): 22, 47, 65, 89, 94, 105, 114, 154, 179
confluent point (of signal paths): 225, 230, 233, 236, 237, 243, 253, 256, 259, 260, 262, 263, 264
conservation (of energy): 17, 18, 20
correspondence principle (of Bohr): 193, 196
decay (of RZ pulse): 151, Section 2.17, 2.18
decision threshold (of logic level): 36, 106, 109, 110, 130
discretization (of time and space): 206, 225, 260
distance (between states): 241
dominant period (of ringoscillator): 244, 245, 246
dynamic switching (of CMOS gate): 34, 40, 55, 198, 199, 269
eigenvalue and eigenfunction: Section 2.04, 2.05
energy transfer (from node to node): 13-16
equation of motion (of excitation): Section 1.07
excitation: Section 1.01 - 1.07, 1.11, 1.12, 1.14, 1.15, 2.01 - 2.03, 2.05, 2.09, 2.11 - 2.13, 2.17, 2.19, 2.20, 3.01 - 3.06, 3.09 - 3.11, 3.14, 3.16, 3.17.
feedback: 9, 54, 78, 99, 100, 103, 111, 112, 127, 129, 138, 141, 152, 158, 189, 192, 196, 197, 201, 210
feedforward: Section 1.14
Fermion (circuit model): Section 2.20
fine structure constant: 17-18
gain-bandwidth product: 119, 153
gate field: Sections 1.01 - 1.07, 1.11 - 1.15, 2.01 - 2.13, 2.17 - 2.20, 3.01 - 3.06, 3.08.
gear mode (of ringoscillator): 234, 244
generation (of transition edge): 57-58, 149-151
gradual-channel, low-field model (of FET): Section 1.04, 1.05, 1.06, 2.05, 2.08, 3.02, 3.03, 3.06, 3.14.
hazard: 57, 58, 148-149, 151, 266
hidden variable interpretation: 13, 106, 109, 271
inductance: 18-20
initial condition (as the source of pulse decay): 165, 172-173
interference (of digital signals): 149

latch (as a decision maker): Section 2.06, 2.07, 2.09, 2.10, 2.12, 2.14, 2.15, 2.16
master-slave data transfer: 148, 155
memory: 129, 155, 158, 202, 204, 205, 206, 258, 259
metastability: Section 2.10, 139, 144, 267
microstates: 203-204
Miller effect: 67, 68, 69, 70, 71, 72, 73, 74
mode (of ringoscillation): 244-248
negative resistance diode: 79-80
neighborhood (of state): 239, 240
neuron: 1, 205, 224, 256
noise: 113
overlap current: 34-35
particle (model of digital excitation): 2. 5, 11, 12, 14, 17, 39, 42, 59, 61, 86, 120, 134, 152, 178, 184, 187, 189, 190, 191, 193, 201, 203, 210, 270
phase-shift oscillator: 195
phase transition: Section 1.12, 178-179, 202
Planck's constant (circuit equivalent): 17, 18
power supply (as the circuit background): Section 1.04, 2.11, 2.12
probability (of Boolean level): 114-116
quantum mechanics: 1, 13, 17, 51-52, 61-62, 84, 86, 106, 109, 112, 117, 121, 124, 130, 134, 189, 190, 191, 194, 196, 203, 210, 270, 271
ringoscillator: Section 3.01, 3.02, 3.03, 3.04, 3.06, 3.07, 3.08, 3.09, 3.10, 3.11, 3.12, 3.13, 3.14, 3.15.
state sequence: 251-256
state space (of ringoscillator): 248-251
static switching (of CMOS gate): 33-34, 47, 269
sense amplifier (as quantum mechanical test equipment): 129
switching threshold voltage (of gates): 3, 26, 51, 54, 55, 56, 62, 90, 101, 102, 105, 106, 107, 108, 109, 110, 112, 117, 121, 123, 124, 132, 155, 165, 171, 172, 197, 199, 222
synchronous mode (of ringoscillation): 239, 241, 245, 254
ternary (buffer and latch): 100
test equipment (of Boolean level): Section 2.06, 2.07, 2.09, 2.10, 2.12, 3.04.
tunnel effect: Section 2.12
uncertainty principle (of Heisenberg, the circuit model): 17
vector gate field: 8-9
velocity (of digital excitation): 5, 7, Section 1.06, 1.07, 75, 81, 132, 159, 160, 181, 200, 201, 212